THE
Practical Surveyor,
OR, THE
Art of *Land-Meaſuring*,
Made E A S Y.

By *SAMUEL WYLD*, Gent.

WITH
NOTES

By *DAVID MANTHEY*

Flower-de-Luce Books
The Invisible College Press, LLC
Arlington, Virginia

The Practical Surveyor,
or, the Art of Land-Meafuring Made Eafy

by Samuel Wyld

Original publifhed in London, England, 1725

Flower-de-Luce Edition
Editing and Notes by David Manthey
Copyright ©2001 by David Manthey

ISBN: 1-931468-06-0

First Printing

Flower-de-Luce Books
    Published by
The Invisible College Press, LLC
P.O.Box 209
Woodbridge, Virginia 22194-0209
http://www.invisipress.com

Please send questions and comments to
editor@invisipress.com

With many thanks to Reb, without whom
this reprint would not have occurred.

Sison's. *Theodolite*

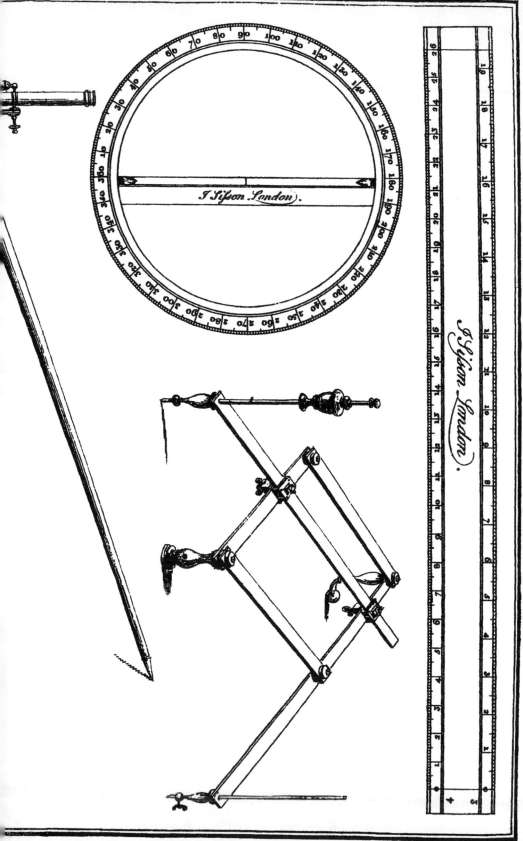

J Sison London).

J Sison London).

THE
Practical Surveyor,

OR, THE

Art of *Land-Measuring*,

Made EASY.

Shewing by plain and practical Rules, how to Survey any Piece of Land whatfoever, by the Plain-Table, Theodolite, or Circumferentor: Or, by the Chain only. And how to Protract, Caft up, Reduce, and Divide the fame.

LIKEWISE,

A New Method of Protracting Obfervations made with the Meridian; and how to caft up the Content of any Plot of Land, by Reducing any Multangular Figure to one Triangle: Being more exact and expeditious than heretofore ufed.

To which is added,

An APPENDIX,

Shewing how to Draw Buildings, &c. in Perfpective, from Obfervations made by the New Theodolite, its Ufe in Levelling, in finding the exact Number of folid Feet, contained in any Timber Trees before they are cut down, by Infpection only; and alfo the Ufe of a new-invented *Spirit-Level*. With feveral other Things never before made Publick.

By *SAMUEL WYLD*, Gent.

LONDON:

Printed for J. HOOKE, at the *Flower-de-luce* againft St. *Dunftan's* Church in *Fleet-ftreet*: And J. SISSON, Mathematical Inftrument-maker, the Corner of *Beaufort-Buildings* in the *Strand*. M.DCC.XXV.

(Price Three Shillings.)

THE
PREFACE
TO THE
READER.

 N this fmall Tract you'll find the whole Art of Surveying Land Epitomized: The Rules and Methods here laid down in a plain and familiar Manner, being fuch as are fitteft for a Practifer's Ufe, without an unneceffary Mixture of ufelefs Curiofities and needlefs Repetitions. And altho' Brevity be chiefly intended, yet nothing is here omitted, but what might well enough be fpared in a Treatife that immediately relates to Practice.

I know

I know the common Objection will be raifed by the Ignorant; that is, What needs any more Writing in this Kind, fince fo many Authors have, with great Ingenuity, beftowed no fmall pains therein; (to whofe Labours and Induftry I acknowledge this Tract not a little beholding.) Now to this Objection, the old and common Anfwer muft be returned, That *A Pigmy mounted on the Shoulders of a Giant, may fee further than its Supporter.* And Arts Mathematical can never be fo fully learned, but that there will ftill be new Experiments left for the Trial of others to fucceed.

Befides, here are inferted, not only the moft ufeful and Practical Methods yet extant in any other Author, but alfo a great many new Improvements never before made publick, rendering the Bufinefs of Surveying Land more exact, eafy and expeditious.

The Book is divided into Seven Chapters, and the Appendix into Three, and thefe into feveral Sections, for the more orderly Ranging the feveral Subjects under their proper Heads.

In the Firft and Second Chapters, is defcribed the Manner of Meafuring Land by the more ufeful inftruments, the Plain Table, Theodolite and Circumferentor; the firft being proper for Gardens, or fuch fmall Pieces of Land

about

about Buildings, the fecond for larger Tracts of enclofed Land, and the third for Parks, Commons, &c.

In Handling this, I have not chofe the moft Accurate Method I could think of, but rather the moft Plain and Simple, as being moft Agreeable to the Conception of a Stranger to the Art, to whom nothing can be too plain.

In the Third Chapter, is defcribed a new Way of Protracting Obfervations made in the field, by the Needle: As alfo how to Caft up the Content of a Piece of Land, by Methods more facile and expeditious than heretofore ufed.

In the Fourth, the Ufe of the Theodolite is fhewn in Surveying feveral Parcels of Land lying together; with the Form of the Field-Book, and Plan of the Work annexed; which fhews, by Infpection only, the feveral Stations and Station-Lines throughout the whole; from which the Obfervations are made in the Field.

The Fifth fhews how to meafure or plot any Piece of Land by the Chain only, without the Help of any other Inftrument in the Field but a fmall Crofs.

The Sixth fhews the Manner of Laying-out, and Dividing Land, without inferting the
various

various methods by which the fame might be performed; but by fuch only as are moft eafy and fit for Practice.

The Seventh fhews how a County is to be Surveyed; as alfo Roads and Rivers; and how to make the Ground-Plott of a City, &c. And becaufe thefe more feldom come in Practice, I only touched generally on the Manner how they are to be performed.

In the Appendix is defcribed the Ufe of a new Spirit-Level, for Conveying Water to any appointed place: Shewing alfo, how the Draught of a Building or other Objects, may be drawn in Perfpective, from Obfervations made with the Theodolite, by a Method entirely new: As alfo how to find a true Meridian-Line, &c. Thefe Problems, tho' not immediately related to the Bufinefs of Land-meafuring, will be found very ufeful to a Practitioner in that Art, and may well deferve the little Room that is allotted them in this Book.

It may be expected, here fhould have been inferted (as ufual in Books of Surveying) more *Theorems, &c.* of Geometry; I confefs it is neceffary a Surveyor fhould be well acquainted therewith, as alfo with Trigonometry, as the Ground-work of the reft. But then he may as well read in the Commentators on *Euclid,* the Demonftration

Demonſtration of each Theorem at large (beginning with the Principals of the Art firſt) as to ſee 'em tranſcribed by Piecemeal any where elſe. Since the Two Theorems in the Firſt Chapter, well underſtood and applied, will be ſufficient for the Performance of moſt Problems relating to Land-meaſuring: And indeed, a Perſon who is well acquainted with the Uſe of his Inſtruments, will have little Occaſion to have Recourſe to Trigonometrical Calculations for finding his Angles, and for caſting up the Content of any Piece of Land after the Plott thereof is made; the Directions in Sect. 4. may be ſufficient: But if any one thinks otherwiſe, he may be farther informed from the Works of our Trigonometrical Writers, of which there are many good ones extant.

However, 'tis hop'd, the Country Farmer, who underſtands but ſo much of Arithmetick, as to add, ſubtract, multiply and divide (with a little Practice, the genuine Parent of Perfection) by theſe plain Directions, and with good Inſtruments, will be enabled to find the Content of each Piece of Land in his own Occupation, (and thoſe who will not be at the Charge of Inſtruments, may make good Uſe of the Fifth Chapter,) and that this Knowledge is extremely neceſſary to the Countryman, none but the

groſſly

grofsly ignorant will deny, fince thereby he may judge what Stock of Cattle each Field will be likely to feed, or what Quantity of Seed will be fufficient for each Acre, or what Number of Workmen to reap or mow the fame, &c. Which makes me admire, when I reflect, that this Science fo beneficial to the Publick, as well as particular Perfons, fhould be fo much neglected, being fo plain and obvious to every Capacity.

But I fhall forbear any Panegyrical Expreffions in Praife of the Art itfelf, (tho' much might be faid on that Head), on Account of its Antiquity, Salubrity, Pleafantnefs, and above all, its Ufefulnefs, *Ornari res ipfa negat contenta doceri.*

As for the Book itfelf, tho' perhaps fome ill-natur'd Artifts may be offended therewith, becaufe feveral Things herein are difcovered (which they would have been as well pleafed fhould have been concealed) like Flowers gathered and placed in one Garland, and proftituted to every one's View; yet if it proves in any way ufeful to thofe for whom it was defigned, I have my End in Publifhing it. *Rumpatur quifquis rumpitur invidiâ.*

Indeed I hoped fome Perfon who had more Hours of Leifure to fpare than my felf, might have fpent fome of them in Compofing
something

ſomething of this Nature, ſince all the Books I
have yet ſeen are much deficient in many of the
moſt neceſſary Parts of the Buſineſs, or elſe too
voluminous for common Uſe; but could not
hear of any ſuch, till this was in the Preſs, and
ſeveral Sheets wrought off, elſe I ſhould gladly
have reſigned the Task: But now that it is
printed, e'en let it be publiſhed; and at the
Bookſeller's Requeſt, I have added thus much
by Way of Preface, which I ſhall conclude with
a Saying I have ſomewhere met with, *Va, mon
Enfant, prend ta Fortune*.

S.W.

THE
CONTENTS.

OW to make a Plott of a Piece of Land by the plain Table, and caſt up the Content thereof.

Sect. 1. *The Quantity of Superficies in an Acre of Land. Tables of Meaſures. Names of Inſtruments.*

Sect. 2. *Directions for Meaſuring Lines in the Field, with the Chain.*

Sect. 3. *How to make a Plott of one or ſeveral Fields together, upon the Paper on the Plain Table, by placing the Inſtrument at one or more Stations about the Middle, from where the Angles may be ſeen.*

Sect. 4. *Directions for Caſting up the Content of any Piece of Land.*

Sect. 5. *How to make the Plott of any Field or Encloſure, on the Paper on the Plain Table, by going round the ſame, and taking Offsets to the Bounders, &c.*

CHAP. II.

CHAP. III.

CHAP.

CHAP. VI.

CHAP. VII.

APPEN-

APPENDIX

CHAP.

CHAP. III.

ADVER-

ADVERTISEMENT.

BECAUSE the Practice of SURVEYING depends much on the *Instruments* used therein, which, being ill-contrived and adjusted, cause unavoidable Errors: Therefore I thought fit to say, That all Sorts of Instruments for Surveying Land, are made with the greatest Accuracy and newest Improvements,

By JONATHAN SISSON,

Mathematical Instrument-Maker,

At the Corner of *Beaufort-Buildings*, In the *Strand*, *LONDON:*

He being the *Only Person* that makes the *Theodolite*, *Spirit-Level*, and *Parallelogram*, hereafter mention'd. Where also any Gentleman or Others may be furnish'd with Sun-Dials of all Sizes, to be fix'd for particular Latitudes, or portable and universal ones, Double Horizontal Dials, and Projections on the Plane of any Circle, Mr. *Collins*'s Quadrants in Brafs, Twelve or Six Inches Radius, Spheres of all Sorts, and Globes, the best Extant: With all other Mathematical Instruments, both for Sea and Land, made in Silver, Brafs, Ivory or Wood: And Sold at Reasonable Rates.

THE
Practical SURVEYOR:

CHAP. I.

Shewing how to make a Plott of a Piece of Land by the Plain Table, and caſt up the Contents thereof.

SECT. I.

Geometrical Superficies or Surface, is produc'd or form'd by the Motion of a Line, as that is deſcrib'd by the Motion of a Point, for if *a b*, in *Fig.* 1. be equally mov'd upon the ſame Plane to *c d*, then will the Points at *a* and *b*, deſcribe the two Lines *a c*, and *b d*; and by ſo doing, they will generate the Superficies of Figure *a b c d*, being the Quantity of two Dimenſions, *viz.* it hath Length and Breadth (but not Thickneſs,) conſequently the Bounds or Limits of a Superficies are Lines; ſo if the Line *a b*, doth contain in Length five Chains, and the Line

Line *a c* two Chains, and if their oppofite Sides and Angles be equal, the Quantity of Land thefe four Lines enclofe, *viz.* (*a b, b d, d c,* and *c a,*) will be an Acre.

Our prefent Bufinefs therefore, will be to compute what Number of fuch Acres, or parts of an Acre, are contain'd in any Piece of Land, (be it Arable, Meadow, or Woodland) whofe Extent is limitted by certain Lines or Bounders. Now an Acre of Land (by the Statute of 33. *of Ed.* I.) is appointed to contain 160 fquare Perches or Poles; it is no matter in what Form it lyes, fo it contains 160 fquare Poles; and *Gunter's* Chain (the beft for Practice) being therefore made four Pole long, ten of thefes fquare Chains make an Acre, (that is to fay) one Chain in Breadth, and ten in Length, or two in Breadth, and five in Length, *&c.* do contain 160 fquare Poles, as *per* Statute. See the following Tables.

I *Table of long Meafure.*

Inches	Link	Foot	Yard	Perch	Chain	Mile
Inches	7.92	12	36	198	792	63360
	Links	1.515	4.56	25	100	8000
		Feet	3	16.5	66	5280
			Yard	5.5	22	1760
				Perch	4	320
					Chain	80

Length of an Acre — Chains	Breadth an Acre — Chains	Links	Pts. of a Link
1	10	00	
2	5	00	
3	3	33	333
4	2	50	
5	2	00	
6	1	66	666
7	1	42	285
8	1	25	
9	1	11	111

A Table

A Table of Square Measure.

	Inch	Links	Feet	Yards	Poles	Chains	Acre
Inch	1						
Links	62.726	1					
Feet	144	2.295	1				
Yards	1296	20.661	9	1			
Poles	39204	625	272.25	30.25	1		
Chains	627264	10000	4356	484	16	1	
Acre	6272640	100000	43560	4840	160	10	1

A Table, shewing how many Chains, Links, and Parts, are contain'd in any Number of Feet, from 1 to 10000.

Feet	Chains	Links	Pts.ofLinks	Feet	Chains	Links	Pts.ofLinks
1	0	1	515				
2	0	3	030	200	3	03	030
3	0	4	545	300	4	54	545
4	0	6	060	400	6	06	060
5	0	7	575	500	7	57	575
6	0	9	090	600	9	09	090
7	0	10	606	700	10	60	606
8	0	12	121	800	12	12	121
9	0	13	636	900	13	63	636
10	0	15	151	1000	15	15	151
20	0	30	303	2000	30	30	303
30	0	45	454	3000	45	45	454
40	0	60	606	4000	60	60	606
50	0	75	757	5000	75	75	757
60	0	90	909	6000	90	90	909
70	1	06	060	7000	106	06	060
80	1	21	212	8000	121	21	212
90	1	36	363	9000	136	36	363
100	1	51	515	10000	151	51	515

The Use of these Tables is plain by Inspection, therefore particular Directions are needless.

Let

Let Figure 2 be fuppos'd to reprefent a Piece of Land, bounded with the four ftrait Lines *a b*, *b d*, *d c*, and *c a*, whofe Lengths are each ten Chain, then the Area or Superficies thereof will contain ten times ten fquare Chains, or ten Acres, (as *per Fig.*) each of the fmall Squares reprefenting one fquare Chain.

But before the Plan of any Piece of Land can be laid down (or protracted) on Paper, in order to make a Computation of the Area or Quantity of Superficies it contains, 'tis neceffary we fhould know the Length and Pofition of the feveral Lines that bound the fame.

And to find the Length and Pofition of Lines in the Field, we make ufe of Inftruments, *viz.*

To meafure the Length of Lines in the Field, we ufe *Gunter*'s Chain, containing in Length four Poles or 66 Feet, divided into 100 Parts or Links, each Link being 7 Inches, and $\frac{92}{100}$ of an Inch, and a Staff whofe Length is equal to $\frac{1}{10}$ part of the Chain; that is to fay, 10 Links, or 6 Foot 7 Inches $\frac{2}{10}$ of an Inch. For Roads the Wheel.

Inftruments us'd for taking the Pofition of Lines, are of two Kinds. With fome we take the Pofition of a Line by the Angle which it makes with the Meridian, ufing a Box and Needle, as the Circumferentor, *&c.* and this is ufually call'd the Bearing of the Line.

With others we take the Pofition by the Angle that the Line makes with any other given Pofition; as with the Limb of the Theodolite, the Chain, *&c.*

But with fome, we take the Angle it felf as with the Plain Table, *&c.*

All other Inftruments either differ from thefe only in Name, or are contain'd in them.

The inftruments for Plotting, are a Scale and
Compafs,

Compass, or rather a Scale decimally divided close to the Edge, and at every tenth Division, numbred 0, 1, 2, 3, &c. denoting Chains, and a Protractor always to be divided, numbred, and fitted to the Instrument.

Of the use of these several Instruments in their Order; and first of the Plain Table; the Uses thereof being as plain as the Name of the Instrument denotes.

But because we make use of the Chain in all manner of Business in the Field, it will be necessary, in the first place, to inform our selves in the manner how to manage it in measuring the Length of Lines in the Field.

SECT. II.

Directions for measuring with the Chain.

The Chain contains in Length 4 Pole or 66 Feet, divided into 100 Links, each Link being $7\frac{92}{100}$ Inches, as aforesaid, having a large Ring exactly in the middle of the Chain, and pretty large Pieces of Brass of different Shapes at the end of each 10 Links, for the speedier counting of the odd Links; also you may tie a large red Rag at 50 Links, and others of a lighter Colour at 25, from each end of the Chain, especially when the Grass is long.

Take care that they who carry the Chain deviate not from a strait Line, (that being the nearest Distance that can be between any two Places,) for if the Lines be not exactly measured, neither the Form nor Content of the Plott can be true. Provide a Staff just six Foot seven Inches and $\frac{2}{10}$ long, which divide into 10 equal Parts, so will the Whole be the Length of 10 Links, and each Part the Length of one Link, and 10 times
the

the Length of this Staff (which you may call the off-fet Staff) the Length of the whole Chain; alfo provide 9 Arrows or fmall Sticks above a Foot long, which you may mark at the Tops with Bits of red Cloath, and at the Bottom you may put fmall Iron Ferrills and two ftrait Staves about 5 Foot each.

Before you meafure with the Chain, 'tis neceffary to examine its Length by the off-fet Staff, ftretching it on level Ground in fuch fort, as when you meafure with it.

Being thus provided, let the Leader of the Chain take the nine Arrows in his Hand, and one of the 5 Foot Staves, and let the Follower, ftanding at the Station, direct him to place his Staff at the Chain's End, in a right Line with the Stations, and there to put down one of his Arrows, and then go on.

Let the Follower, being come to the Arrow, take it up, and put his Staff in the place thereof, and direct the Leader to place his Staff as before; then let the Leader, ftanding at his Staff, look back towards the laft Station, and he will fee the two Staves and the Station in one right Line, if they have directed right; but if not, the Leader muft direct the Follower to place his Staff in the fame right Line with the Station and the Leader's Staff, and fo muft each direct the other, till the two Staves and two Stations are in one right Line.

Let 'em thus proceed till they have meafured to the Station, or till the Leader is nearer the Station than one Chain's Length; then will the Number of whole Chains meafured, be exprefs'd by the Number of Arrows prick'd down, fuppofe 7; and the Leader holding the End of the Chain to the Station, and the Follower will fee how

many

many Links are contain'd between the Station and laſt Arrow, ſuppoſe 60.

Now enter in the Field Book, the Chains and Links without any Diſtinction between them, and they will be 760 implying either 760 Links, or 7 Chains 60 Links; but the Links muſt always poſſeſs two Places after the Chains, as 7 Chains 4 Links muſt be written 704, and not 74; and 8 Chains muſt be written 800, alſo as ſoon as you have meaſur'd each Length, enter it down immediately in the Field Book, and never truſt to your memory; alſo at the end of each Length, inquire of the Leader and Follower how many Arrows each have in their Hands, and if the Sum of the Arrows are not nine, the laſt meaſur'd Length is doubtful, and muſt be re-meaſur'd before you proceed.

When the Length is above nine Chains, let the Leader go on, and ſet his Staff down at the tenth, and let the Follower put his Staff in the place of the Leader's, and give the Leader the nine Arrows, and then proceed as before; but enter theſe 10 Chains immediately in the Field Book, and if the Length is 10 Chains more, enter 20; alſo obſerve it is uſual to allow 5 Links from the Stem of the quickſet Hedge, for the Breadth of the Ditch, except the Cuſtom or Agreement is otherwiſe; but the Cuſtom of the Place generally is the Surveyor's Rule.

S E C T.

SECT. III.

Shewing how to make a Plott of one Field, or several Fields, upon the Paper on the Plain Table, by Placing the Instrument at one or more Stations about the Middle, from whence the Angles may be seen.

Let *Fig.* 3. be suppos'd to reprefent two Fields or Enclofures, a Plott of which is defir'd, and firſt of the Field, *a l m o b.*

Having put your Plain Table in order, and obſerv'd the Needle to play well, put a Sheet of fair Paper thereon, and cruſh down the Frame, ſo that the Paper lyes ſmooth; then you may imagine the Paper on the Table to reprefent the Surface of the Land, and the Lines you ſhall draw thereon, to be the Bounders of the reſpective Fields or Enclofures in ſome Proportion or other, and if you make the Proportion thus, A Line drawn on the Paper an Inch long, is equal to, or in proportion to, the Length of one Chain on the Land, and if 5 Chains in Length, and 2 in Breadth, do contain the Quantity of one Acre on the Land, then 5 Inches in Length, and 2 in Breadth, ſhall alſo contain the Quantity of one Acre on the Paper, and takes up the ſame Quantity of Space thereon, as an Acre does in the Field; *vide Fig.* 1.

This being premis'd, we proceed to lay down the Lines that enclofe thefe two Fields of the fame Length and Pofition on the Paper, which ſhall therefore include the fame Quantity of Superficies as thofe in the Field it felf in proportion, as 1 Inch to 1 Chain.

But if we make half, or a quarter of an Inch, or half a quarter (by which the following

Dimensions were laid down) on the Paper, to represent one Chain, it is the same in effect, only the Plott thereby is made lesser; also the Inches, &c. on the Paper, are capable of being subdivided, as the Chain is into Links.

First, place the Table somewhere about the middle of the Field, from whence, if possible, you can see all the Angles, as at ☉, and make an Hole in the Ground, over which by the help of a Plummet and String, set the Center of the Table, by applying the String to the head of the Staff, and having set the Instrument steady on the Ground, turn the Table about till the Needle hangs over the Flower de Luce in the Box, (or if 'tis more convenient, turn the Length of the Table to the Length of the Plott, and note, on a bit of waste Paper, the Division in the Box the Needle hangs over when at rest,) and then screw the Table fast. Assign on the Paper a Point, or stick a Pin at ☉, (to represent the Hole in the Ground or present Station) to which Point, apply the fiducial Edge of the Index, and turn it about, keeping the Edge close to the Point or Pin at ☉, till through the Sights you see the Hair cut a Staff or Mark, set up exactly in one of the Angles, as at *a*, then by the Edge of the Index, draw a Line from the Point ☉ towards the Angle *a*, with the Point of the Compasses, without regarding the Length, so it be long enough.

Let the Mark be left at *a*, and cause others to be set up round the Field, at every Angle therein, as at *l m o b*, to every one of which direct the Sights, and when the Hair therein cuts the Mark (keeping the Edge of the Index close to the Point ☉) draw the several Lines ☉ *l*, ☉ *m*, ☉ *o*, ☉ *b*.

Now see whether the Needle continues to
hang

hang over the fame Point in the Box as when you firft planted the Table, alfo lay the Edge of the Index to the Line ☉ *a*, and if the Hair in the Sights cuts the Mark at *a*, and the Needle hangs over the fame Point as at firft, you may conclude the Table hath not been mov'd out of its firft Pofition, which is carefully to be obferv'd.

In the next place we proceed to make the Lines ☉ *a*, ☉ *l*, &c. of their juft Length, and to that end apply the Ring at one end of the Chain to the Hole under the Table, and let the Chain be ftretch'd at length towards one of the Angles as at *a*, and when I have meafured up thereto, (obferving the Directions before laid down for meafuring with the Chain,) I find the Length of the Line ☉ *a*, to contain 3 Chains 60 Links, or 360 Links, which I note in a Bit of Paper.

Having found the Length of the Line ☉ *a* on the Ground, I proceed to make that on the Paper of the fame Length, to correfpond therewith in Proportion, as an Inch to a Chain, or any other Proportion as fhall be moft convenient.

Therefore having provided a Scale and Compaffes, becaufe the Length of the Line I am about to take off, is 360, or 3 Chains 60 Links: I fet one Foot of the Compaffes in the Line of Inches, &c. at 3, and extend the other to 60 in the diagonal Divifions; thereby taking off 360 Links, then the Compaffes remaining at this Extant, I fet one Foot in the Point ☉ on the Paper, and let the other fall in the Line ☉ *a*, where I make a vifible Mark or Prick with the Compafs Point, in order to find it again prefently.

Note, each whole Inch, ½ Inch, &c. on the Scale, reprefents one Chain, and the Links, or hundredth Parts of the Chain, are taken off from one whole Inch, ½ Inch, &c. divided alfo into 100

Parts,

Parts, by diagonal Lines drawn crofs the Scale; for each 10*th* of that Divifion reprefents 10 Links, and is divided into other ten Parts, by the croffing of the diagonal Lines. Any Inftrument-maker that fells this Scale, will prefently fhew the Ufes of thefe Lines, or the Sight thereof will be fufficient Information; only it may be obferv'd, that thefe Scales are beft made of Brafs, and the Joints of the Compaffes are chiefly to be obferv'd, which fhould have an equal eafy Motion without leaping, and that the Points be well temper'd, and clofe in a Point exactly.

Obferving the Directions before laid down, I firft meafure with the Chain the Length of the feveral Lines ☉ *l*, ☉ *m*, ☉ *o*, and ☉ *b*, and then tranffer the Length of each Line on the Ground to its Reprefentative on the Paper, making vifible Marks where the end of each Line falls, as at *a*, *l*, *m*, *o*, *b*.

And here it may be obferv'd, that I generally meafure one Line from the Inftrument to an Angle, and the next from the Angle to the Inftrument, till all are finifh'd, and note the Length of each Line as I meafure it in a Piece of wafte Paper, and never truft to my Memory; then obferving which Line I began with, I fet on its true Length on the Paper on the Table, and the reft of the Lines in their Order.

Laftly, I join the Points *a*, *l*, *m*, *o*, *b*, with Ink Lines, becaufe they fhould not rub off, (and for this purpofe a drawing Pen is requifite,) as the Lines *a l*, *l m*, *m o*, *o b* and *b a*, which conftitute the Bounders of the Field *a l m o b*.

It may be a young Beginner will take fome Pleafure to meafure crofs fome part of the Plott on the Paper, as the Diftance from *a* to *o*, or from *a* to *b*, with his Scale and Compaffes, and afterwards
meafuring

meafuring the fame Diſtance on the Ground with the Chain, he will find them both exactly to agree, if the Plott be truly laid down.

Having finiſh'd this Field, I cauſe as Staff to be ſet up with a Paper thereon in the next, in a convenient Place, from whence I can view all the Angles; but if ſuch a Station can't be found, I choſe the moſt convenient as at ☉ 2.

The Table ſtanding at ☉ 1, in the Field *a l m o b*, in the fame Poſition as at firſt (which it muſt do, or the Plott of the next Field can't be truly laid down in reſpect of the laſt) I lay the Index to the Point ☉, and turn it about thereon, till the Hair in the Sights cuts the Staff or Mark in the next Field at ☉ 2, and holding the Index faſt in that Poſition, I draw a Line by the Edge thereof, from ☉ 1, towards ☉ 2, in the next Field, and take care to continue it long enough; then remove the Table, and place a Staff with a Mark thereon, in the Hole over which the Center of the Table was plac'd, and meaſure with the Chain the neareſt Diſtance between ☉ 1, (in the Field *a l m o b*,) and ☉ 2 in the next Field, (drawing the Chain through the Hedge in a ſtrait Line, and ſet on the Diſtance 621, (by the Help of the Scale and Compaſſes) from ☉ 1 to ☉ 2.

Now I take away the Staff, and plant the Center of the Table over the Hole, in which the Staff ſtood at ☉ 2, and ſticking two Pins, or the Points of two ſmall Needles in ☉ 1, and ☉ 2, apply the Edge of the Index thereto, ſo as it lyes exactly on the Line ☉ 1, ☉ 2, and keeping it in this Poſition, turn the Table about till the Hair or Thread in the Sights cuts the Staff or Mark in the laſt Field, and then ſcrew the Table faſt that it ſtir not out of this Poſition, till I have finiſh'd the Obſervations in this Field, (but obſerve to turn that Part of the Table mark'd ☉ 1, towards its Repreſentative in the laſt Field.)

Field.)

When the Needle hath fettled, and is at reft, obferve whether it hangs over the Flower-de-luce or fame Divifion in the Box, as at ʘ 1 in the laft Field; which it will do if you have made your Obfervation truly, and the Needle be good; and if you were to move the Table to never fo many Stations, the Needle will ftill point to the fame Direction in the Box, which obferve; becaufe the removing the Table from one Station to another is the greateft Difficulty in this Way of Surveying.

Having caus'd Marks to be fet up in fo many of the Angles in this Field, as you can conveniently fee, from the prefent Station, as at *k, i, h, g,* lay the Index to the Point ʘ, and direct the Sights to *k, i, h, g,* drawing Lines by the Edge of the Index towards every one of them; then meafuring the Length of the feveral Lines ʘ *k,* ʘ *i,* ʘ *h,* ʘ *g,* with the Chain, fet on the feveral Lengths of thefe Lines on the refpective Lines on the Paper (as before directed,) marking the Points *k, i, h, g,* where the Ends of the Lines fall from ʘ: Laftly, join the Points *l k, k i, i h,* and *h g,* with ink Lines, and they conftitute the Bounders of fo much of the prefent Field, as you can conveniently fee from this Station.

But there is no occafion to meafure the Angles *l* or *m* with the Chain, except that it may be fome Satisfaction, as aforefaid, to fee the Lines on the Paper, and thofe on the Ground to agree.

Obferving the former Directions for removing the Table, let it be plac'd in its true Pofition at ʘ 3 in this Field; and direct the Sights to the Angles *f, e, d,* and *c,* and when the feveral Diftances from ʘ 3, to *f, e, d,* and *c,* are fet on the Paper, join the Points *g f, f e, e d, d c,* and *c b,* with ink Lines, fo is the true Plott of thefe two Fields, *a l m o b,* and *l k i h g f e d c l o m,* laid down on the

Paper

Paper in such Proportion, as the Scale you made use of is to the Chain.

But obferve that if the Hedge *b c*, had been fo thick, that from ☉ 3, you could not have feen the Angle *d*, or other Obftruction had hinder'd your Sight or Meafuring thereto, you muft have remov'd the Table to another Station; but when you can (as commonly you may) by holding afide Boughs or otherwife, fee the Mark, and by drawing the Chain through the Hedge, meafure the Line from ☉ 3, to the Angle *d*, 'tis a better way than to remove the Table, for the fewer Stations you make, the better, and the Work will be more truly laid down, which remember.

It would only be repeating the fame thing over again, to give Directions how to furvey a Field from a Station taken in any Angle thereof, from whence the reft may be feen; as if it had been more convenient in the Field *a l m o b*, to have planted the Table at the Angle *a*, the Sights muft have been directed from thence to the reft of the Angles *l*, *m*, *o*, *b*, and the Lines meafured on the Ground, from *a* to *l*, *m*, *o* and *b*, whofe Length laid down on the Paper from *a*, would give the fame Points *l*, *m*, *o*, *b*, as if the Station had been in the Middle of the Field, and when the Bounders are drawn, they will be in the fame Pofition as before.

If you would draw a Meridian, or a North and South Line through the Plott, turn the Table about, till the Needle hangs over the Flower-de-luce in the Card, and laying the Index on fome of the equal Divifions on the fide of the Frame, draw a Line clofe by the Edge, which fhall be a meridian Line; and if you crofs this Line by another at Right Angles, that fhall fhew the Eaft and Weft Points.

S E C T.

SECT. IV.

Directions for casting up the Content of any Piece of Land.

The next thing that lies before us is the Manner of Calculating the Quantum of the Superficies enclosed by the Lines on the Paper, as they represent the Bounders in the Field; that is to say, how many Acres, and Parts of an Acre and contain'd therein.

In order thereto, these several Things may be premised: First, That every Magnitude is mensurable by some Magnitude of the same Kind; as, a Line by a Lineal Foot, &c. a Superficies by a Square Foot, &c. and a Solid by a Solid Foot, &c. The Superficial Measure may be conceived, by imagining *Fig.* 2. to be a Field, as now divided into 100 Squares, every Square being one Chain, having a Chain for its Side: Now if the Field be just one Chain Broad, the Number of Square Chains will be equal to the Number of Lineal Chains in the Length of the Field: But if the Field be 2, 3, 4, &c. Chains broad, the Number of Square Chains will be twice, thrice, or four times so many square Chains; so this Field being 10 Chains long, and 10 Chains broad, the Number of square Chains in it are, 100, *viz.* 10 multiply'd by 10, gives 100; or if it had been 40 Chains long and 5 broad, the Number of square Chains would be 200.

2. That the Mensuration of all superficial Figures, as Land, &c. depends on the exact Measuring of certain regular Figures, as the Parallelogram, Triangle, &c. so that if any Plot of Land be not one of these Figures, it must be reduc'd into some or one of these Forms before it

can

can be meafured.

3. A Parallelogram is a quadrilateral Figure, each of whofe oppofite Sides are parallel, and the Diagonal divides the fame into two equal Parts, as the Parallelogram *a b c d, Fig. 5.* the oppofite Sides and oppofite Angles of which are equal between themfelves, and the Diameter or Diagonal *a c* bifects the Parallelogram.

4. A Right-lin'd Triangle, is a Figure comprehended within three ftrait Lines.

We need not here take notice whether a Triangle be Right or Oblique-angled, or by what Name diftiguifh'd, whether an Ifofceles, Scalenum, *&c.* becaufe they are all meafured by one and the fame Rule.

5. A Trapezia is an irregular four-fided Figure, comprehended under 4 unequal Sides and Angles.

To this we fhall add thefe two ufeful Theorems following.

Theorem 1.

That Parallelograms conftituted upon the fame Bafe, and between the fame Parallels, are equal, *Euclid. lib.* 1. *Prop.* 35.

Let *a b c d, e b c f,* be Parallelograms conftituted upon the fame Bafe *b c,* and between the fame Parallels *a f,* and *b c,* then the Parallelogram *a b c d,* is equal to the Parallelogram *e b c f.*

For becaufe *a b c d* is a Parallelogram, *a d* is equal to *b c,* and for the fame Reafon *e f* is equal to *b c,* wherefore *a d* fhall be equal to *e f,* but *d e* is common, therefore *a e* is equal to *d f,* but *a b* is equal to *d c,* wherefore *e a, a b,* the two Sides of the Triangle *a b e* are equal to the two Sides *f d, d c,* each to each, and the Angle *f d c,* equal to the

Angle

Angle *e a b*, the outward one to the inward one, therefore the Bafe *e b* is equal to the Bafe *f c*, and the Triangle *e a b*, to the Triangle *f d c*.

If the common Triangle *d g e* be taken from both, there will remain the Trapezia *a b g d*, equal to the Trapezia *f c g e*, and if the Triangle *g b c*, which is common, be added, the Parallelogram *a b c d*, will be equal to the Parallelogram *e b c f*, which was to be demonftrated. *Vide Fig.* 4.

Theorem 2.

If a Parallelogram and a Triangle have the fame Bafe, and are between the fame Parallels, the Parallelogram will be double to the Triangle, *Euclid. lib.* 1. *Prop.* 41.

Let the Parallelogram *a b c d*, *Fig.* 5. and the Triangle *e b c*, have the fame Bafe, and between the fame Parallels *b c*, *a e*, then the Parallelogram *a b c d*, is double the Triangle *e b c*.

For join *a c*, then the Triangle *a b c* is equal to the Triangle *e b c*, for they are both conftituted upon the fame Bafe *b c*, and between the fame Parallels *b c*, *a e*, but the Parallelogram *a b c d*, is double the Triangle *a b c*, fince the Diameter *a c* bifects it, wherefore likewife it fhall be double to the Triangle *e b c*, which was to be demonftrated. *Vide Fig.* 5.

By the laft *Theorem*, a Parallelogram having the fame Bafe with a Triangle, and lying between the fame Parallels, is double to the Triangle; therefore if the Bafe of a Triangle be of the fame Length with one fide of a Parallelogram, and the Perpendicular be of the fame Height, as the other fide of the Parallelogram, thofe Triangles are equal to but half that Parallelogram, *Vide Fig.* 6.

If

If a Weight (as a Bullet) was fufpended at *e*, *Fig.* 6. and from thence let fall on the oppofite Line *b d*, it wou'd defcribe the Perpendicular Line *e f*, then a Perpendicular is the neareft Diftance, or fhorteft Line that can poffibly be drawn from any Angle to its oppofite Side.

And here it may be convenient to infert the Manner of raifing and letting fall a Line perpendicular to another.

Let a Perpendicular be raifed on the Line *d e*, from the Point *c*, *Fig.* 7. Open the Compaffes to a convenient Diftance, and mark out the Points *a* and *b*, then opening them fomething wider, you may be fetting one Foot in *a* and *b*, feverally defcribe the two Arches cutting one another at the Point *f*, from which draw the Perpendicular *f c*, to *c*, alfo the Line *f c*, continued, will crofs the Line *d e*, at Right Angles.

Let a Perpendicular be raifed from the End of the Line *a b*, at *b*, *Fig.* 8. Open the Compaffes to an ordinary Extent, and fetting one Foot in the Point *b*, let the other fall at Adventure, as at ⊙, then without altering the Extent of the Compaffes, fet one Foot in the Point ⊙, and with the other, crofs the Line *a b*, at *d*, and defcribe the obfcure Arch *d d*, then lay the Ruler to *d* and ⊙, and draw the obfcure Line *d* ⊙ *c*, through the Arch. Laftly, apply the Ruler to the Point of Interfection at *f*, and to the Point at the End of the Line *a b*, and draw the Perpendicular *c b*.

Let a Perpendicular be let fall from the Point *c*, upon the Line *a b*: Set one Foot of the Compaffes in the given Point *c*, and with the other, defcribe fuch an Arch of a Circle, as will crofs the given Line *a b*, in two Points, *viz. d, f*, then bifect the Diftance between the two Points *d* and *f*, as at *e*, and draw the Perpendicular *c e*.

This

This is no more but the Firſt Problem re-
verſed: Alſo a Perpendicular may be let fall nigh
the End of a Line by the Second.

Note, Thoſe Problems touching Perpendiculars, aim at no greater Matter, than may be performed in a mechanical Way, by the help of a
ſmall Square, exactly made as a Square Protract-
or, or for the want thereof, a Scale in a Caſe of
Inſtruments, that hath a Right Angle, and true
Sides, or the Square therein, in the Form of a
Carpenter's, *&c.* for if you apply one Leg of ſuch
a Square to any Line, ſo as the Angle of the
Square may touch the End of the ſaid Line, or
any other Point where the Perpendicular is to be
raiſed, you may by the other Leg, draw the
Perpendicular. In like manner may you let fall a
Perpendicular from a Point aſſigned, by applying
one Leg of the Square to the Line, ſo as the
other may touch at the ſame Line the aſſigned
Point, whence you may draw the Perpendicular
by that Leg that touches the Point.

If the Angle of the Square, be a little Blunt by
Wearing, you muſt allow for it when you apply it
to the Point in a Line, and when you are drawing
a Perpendicular, you muſt ſtop before you reach
the given Line, and then by applying the Leg of
your Square, to that Part of the Perpendicular
already drawn, ſo as that Part of the Leg may
paſs clearly over the Line, you may draw the
Perpendicular as exactly, as if the Angle had
been true. The like Courſe is to be taken, when
a Line is to be croſſed by another quite through
it at Right Angles.

6. Every Figure encloſed with 3 Right-lines is
a Triangle, and in the Menſuration thereof, only
the Length of the Baſe, and the Height of the
Perpendicular is conſiderable, and any of the
Lines

Lines may be made the Bafe, though commonly the longeft is, and a Line let fall from the oppofite Angle upon, or made to touch the Bafe in the neareft Point, is the Perpendicular, and you are not confin'd to any Angle, but may let fall the Perpendicular from what Angle you pleafe, taking the Line on which it falls for the Bafe.

7. The Area or Content of the two Primitive Right Lin'd Figures, the Square or Parallelogram and Triangle is found by multiplying the Length of the Square or Parallelogram by its Breadth, and the Bafe of the Triangle by half the Perpendicular, or the whole Perpendicular by half the Bafe, or the whole Bafe by the whole Perpendicular, then half of that Product, is the Content of the Triangle.

And here it may be worth obferving, that the Multiplier in any of the Multiplications made ufe of in cafting up any Mefuration, is an abftract Number, as well as in all other Multiplications whatfoever, which may prevent falfe Confequences ufually drawn from multiplying Feet, &c. by Feet, (*viz.*) that of multiplying by a contract Number (as half a Crown by half a Crown) which is contrary to the Nature of Multiplication, whofe Operations are only compendious Additions, either of the Multiplicands, or fome Part of it continually to its felf or its Part.

8. The Parallelogram or Square, being the original Figure from which are deduced all Computations, that relate to the cafting up the Content of a Superficies, a Line drawn from any Angle therein to its oppofite Angle, may be divided into two Triangles (which dividing Line is called the Diagonal, as aforefaid) both of which
Triangles

Triangles taken together, are equal to the Square or Parallelogram, and one of them equal to half of it, and any manner of Figure that hath four Sides, whether equal or parallel, or neither, let it be called a Trapezia-Rhombus, &c. is capable of being divided and caft up in the fame manner.

9. Then any irregular Figure, let it confift of never fo many Sides, may be divided by fuch Diagonals into a Number of Triangles, &c. which feparately caft up and add together, their Sum is the Content of the whole Figure.

10. And almoft all Fields to be met with in Surveying, being bounded with a Number of unequal Lines, we firft take the Plott thereof by fome Inftrument, and lay it down on Paper, and by drawing Diagonal Lines through it, reduce it into Triangles, &c. *Vide Fig.* 10.

11. And thefe Triangles or Squares, being meafured by the Chain of 100 Links, when caft up, their Content is given in the loweft Denomination, (*viz.*) fquare Links, as in Figure 10: 10 Chains multiplied by 10 Chains, gives 100 fquare Chains; or, which is all one, 1000 Links by 1000 Links, give 1000000 fquare Links in the loweft Denomination, only making the Links to poffefs two Places of Figures after the Chains, as 6 Chains 54 Links muft be written 654, without any Point of Separation between them, and 10 Chains muft be written 1000.

12. In one fquare Chain, there are 10000 fquare Links, and 100000 fquare Links in an Acre; the Chain therefore is divided into 100 Parts or Lengths (fuch as a Link is made to be) on purpofe, that all Operations may be made in a decuple Manner, and to fave the trouble of Divifion: For as the Acre is limitted by Statute,

this

this Number 100000 is the Divifor in the loweft
Denomination (*viz.*) Links. So if a Field contains
1654321 fquare Links, we need not to find the
Number of Acres therein divide it by 100000, the
fquare Links in one Acre; but according to the
old Rule, when a Divifor confifted of 1 and
Cyphers, cut off from the Right-hand, fo many
Places of Figures of the Dividend, as the Divifor
hath Cyphers, accounting them the Remainder;
fo fhall the reft on the left Side be the Quotient;
fo the five laft Figures cut off from the Right,
there are 16 Acres for the Quotient, the
remainder Parts of an Acre. Then by the known
Rules in Decimal Arithmetick (*viz.*) multiplying
Decimal Fractions by known Parts in the next
inferior Denomination, gives thofe known Parts
in Integers; due regard being had to the
Separation. So 54321, multiplied by 4, the Roods
in an Acre, produces 217284 from which Product
cutting off 5 Places of Figures towards the Right,
leaves 2 Roods on the left; and that Remainder
fo cut off, being multiplied by 40, produces
691360; and from this laft Product, feparating 5
Places by a Point, gives 6 Poles on the Left and
the Remainder $\frac{91360}{100000}$ Parts of a Pole.

From what hath been faid, the general Rule
for cafting up the content of a Piece of Land may
be given.

General Rule.

Set down the Number of Chains and Links in
the Order of Multiplication, making the Links
poffes two Places after the Chain; and from the
Product cut off by a feparating Point 5 Figures to
the Right-hand, fo fhall thofe on the left be
Acres: Then multiply the five Figures fo cut off
by

by 4, (the Roods in an Acre) feparating five Places alfo from that Product towards the Right Hand; then the Figures on the left of the feparating Point are Roods.

Laftly, Multiply this laft Remainder by 40, (the Poles in a Rood) and feparate five Places to the Right from that Product, and the Figures on the Left fhow the Number of Poles.

So in *Fig.* 10. the Field *a l m o b* being divided into the Triangle *l m o*, whofe Bafe is 660, and the Perpendicular let fall from the Angle *m*, on the Bafe Line *l o*, is 252.

Therefore half 600, multiplied by whole 252; or whole 660 multiplied by half 252; or whole 660, by whole 252, then the half of this laft Product is the true Content. Ufe any of thefe Methods, the Content of the Triangle *l m o* will be 83160.

The Diagonal Line *l b* divides the Trapezia *l o b a* into two Triangles, *l o b* and *l b a*, which might be feparately caft up as the Triangle *l m o*; but the quicker Way is, to add the two Perpendiculars *o z* and *a x* together, and by that Sum multiply the Line *l b*, which is a common Bafe to both Triangles, and halve the half Product for the true Content of the Trapezia. See the following Work.

Bafe 660 —— *l o*   Perp. 290 —— *a x*
Perp. 252 —— *m y*   Perp. 272 —— *o z*
        ———   ———
        1320   562   Sum
        3300   Bafe 800 —— *l b*
        1320   ———
     ————   449600
     166320

                166320 ⎱
                449600 ⎰ Product
                ————
                615920   double Content
                ————
Acres —— 3.07960   true Content
                4
                ——
Roods —— .31840
                40
                ——
Poles —— 12.73600

        Acres   Roods   Poles   Parts of a Pole
          3 :   0 :   12 : $\frac{73600}{100000}$

The Bafe *l o* multiplied by the Perpendicular *m y*, produces 166320, which is double the Content of the Triangle *l m o* in fquare Links. Alfo the Perpendicular *a x*, added to the Perpendicular *o z*, makes the Sum of both 562, which multiplied by 800, the common Bafe to both Perpendiculars, produces 449600, which is double the Content of the Trapezia *l o b a* in fquare Links.

Therefore (for avoiding Fractions) the double Content of the Triangle *l m o* 166320, added to the double Content of the Trapezia *l o b a*, 449600 gives 615920 the double Content of the
Field

Field *a l m o b* in fquare Links, the half of which (*viz.*) 307960 is the true Content of the Field *a l m o b* in fquare Links, which reduced into Acres, &c. as before directed gives 3 Acres, 12 Poles, and a little above half a Pole, for the true Content of the Field *a l m o b*; but the Parts of a Pole are feldom regarded.

In the fame manner the Field *b c d e f g h i k l m o*, *Fig.* 10. being divided into Trapezias and Triangles, add both the Perpendiculars of each Trapezia together, and by that Sum multiply the Diagonal or Bafe: Alfo multiply the Bafe of each Triangle by the Perpendicular, and fet the Product of each Trapezia and Triangle in an or-derly manner, one under another, and add them all together into one Sum, the half of which Sum is the Content of the Field in fquare Links, which reduce to Acres, &c. as aforefaid.

But remember to meafure the Bafes and Per-pendiculars by the fame Scale that the Plott was laid down by, and contrive to reduce the Field into as large Trapezias and Triangles as poffible, for the fewer you make, the exacter will the Work be caft up; and draw the Bafe-lines neat and fmall, and exactly from Angle to Angle, and let the Perpendiculars juft touch the Line, but not pafs over it, at the neareft Diftance from the Angle that may be; and for this End a good pair of Compaffes, and a Diagonal Scale are moft proper; and the larger the Scale you ufe the better, if the Compafs of the Plott will admit thereof.

S E C T.

SECT. V.

Shewing how to make the Plott of any Field or Enclosure, on the Paper on the Plain Table, by going round the same, and taking Offsets to the Bounders, &c.

Let *Fig.* 11. represent a Field to be plotted by the Plain Table.

Now the former Method of planting the Table at one Station or more, in the middle of the Field, and measuring from thence the Distance to every Angle, is easiest for a Beginner, but is not convenient in many Cases; because he may be hindred by Firze, Water, *&c.* from measuring the Lines to all the Angles; and in many Fields where the Fences are as irregular as the Side *a e* in this Field, he will be obliged to measure a great Number of such Lines.

'Tis therefore best to plant the Instrument at the most remarkable Angles, and measure round the Field, for by this Method, all Sorts of Land may be measured (so the Plan be not too large for one Sheet of Paper,) either within or without the Plott, as Convenience shall determine.

Note, This Mark ⊙, always represents a Station, ... a Prick Line represents the Station Line, and ⸺ a Black Line the Bounder.

First set up a Mark at *a*, and draw a Line on the Table, to represent *a b*, in the Field, then measure the Distance to the Hedge from ⊙ 20 Links, which set from ⊙ to *f*, also measure the Distance from ⊙ *a*, to ⊙ *b*, 840 Links, which set on the Line *a b*.

But instead of using a Scale and Compasses, you may set on the Distance by the Plotting Scale only, such as aforementioned, whose Edge

is

is champered, and the Numbers, and Divisions
set close thereto; (being much readier than
Compasses, and generally used by the best
Surveyors) each Division representing a Chain,
being numbred 1, 2, 3, &c. and each of those
Divisions being again sub-divided into ten Parts,
one of which ten Parts represents ten Links.

Therefore lay the Edge of this Plotting Scale
close to the Line *a b*, the Beginning of the
Numbers coinciding with *a*, and encreasing
towards *b*, and because the Length of the Line *a
b*, is 840, make a Mark with a Needle or
protracting Pin against 840, close to the Edge of
the Scale.

But if the Number had been 845, &c. you must
guess at the odd Links, which you may do by a
Scale of ¼ of an Inch within 2 in an Hundred, and
not mistake, but with a larger Scale much nearer.

Having drawn the Line *a b*, place the Table at
b, and lay the Edge of the Index close to the Line
a b, and turn the Table about till you see the
Mark at *a*, and there screw it fast; then turn the
Line the Index about on *b*, till you see a mark at
c, and draw *b c*, with the Point of the Compasses,
or a Black-lead Pencil; also direct the Sights to the
Barn, and draw the Obscure Line *b z*, not
regarding its Length, so it be long enough.

When the Needle hath settled, take notice
what Division in the Box it points to, for to that
Division it will point at every Station through
the Plott, if your Work be true, and the Needle
good, as aforesaid; but because 'tis not
convenient to trust the Needle when we can do
without it, I shall here lay down a surer way to
correct an Error, before 'tis communicated to the
following Part of the Work.

In the next place, Measure the Distance from

⊙ *b*

⊙ *b*, to the Hedge 17 Links, which set from ⊙ to *g*, and draw the Bounder *f g*, also measure the Distance from ⊙ to *h*, which set on the Paper from ⊙ to *h*, and continue the Bounder *f g*, in a Strait Line, as you see it in the Field.

Remove the Table from *b*, and set up a Staff with Paper thereon in the Hole, over which the Center of the Table stood, as by former Directions, and measure from *b*, towards *c*, with the Chain, but when you come over against the Bend in the Hedge at *i*, measure the Distance from the Chain Line *b c*, to that Bend 7 Links, which set from the Chain Line *b c*, to *i*, and draw the Bounder *i h*, through *h*, till it cuts the Bounder *f g*, constituting that Corner of the Field.

Measure on to *c*, 620 Links, which set from *b* to *c*; now the Reason why we made the Station *b*, so far from the Corner, is to avoid Planting the Instrument too often, for if we had continued the Station Line *a b*, into the Corner, we must have made another Station at *i*, otherwise we could not see to the Angle at *c*, for the fewer Stations we make, the exacter will be the Work, as aforesaid.

Now in order to examine the Length of *b c*, and also its Position in respect of *a b*, do thus: Plant the Instrument at *c*, and lay the Index on the Line *b c*, and by turning the Instrument about, direct the Sights to *b*, and there screw it fast; then turn about the Index on Point *c*, towards the Mark at the Angle *a*, in the Field, and if the Edge does not cut the Point *a*, in the Table, the Line *b c*, is false, either in Position or Length, and therefore must be corrected before you proceed.

The Line *b c*, being truly laid down, and the
Table

Table ftanding at *c*, in the fame Pofition, lay the Index to the Point *c*, and turn it about thereon, till the Hair in the Sights cuts the Mark at *d*, and draw the Line *c d*, alfo direct the Sights to the Middle of the Barn, the Index being turned about on the fame Point *c*, and draw the Obfcure Line *c x*, croffing the other Obfcure Line *b z*, fo fhall the Point of Interfection determine the Situation of the Barn in the Middle of the Field, which you may prove by meafuring on the Ground thereto, from any Part of the Field.

Next meafure the Diftance from the ⊙ at *c*, to the Hedge 6 Links, which fet from ⊙ to *k*, and draw the Bounder *i k*, continuing the Line through *k*, alfo meafure the Diftance to the other hedge *c d*, from ⊙ *c*, 15 Links, which fet off to *l*.

Remove the Table from *c*, and place a Mark there, and meafure the Diftance from *c* to *d*, 481 Links, which Diftance fet on the Line *c d*, then plant the Table at *d*, and lay the Index on *c d*, and turn the Table about till you fee a Mark at *c*, and then fcrew the Inftrument faft.

Next, Examine the Length and Pofition of *c d*, in refpect of *b c*, as before directed, then turning the Index about on *d*, direct the Sights to *e*, and draw the Line *d e*, and fet off the Diftance from ⊙ *d*, to the Hedge at *m* 10 Links, and from *m*, draw the Bounding Line *m l*, continuing it ftreight through *l*, till it croffes *i h*, as you fee it to do it in the Field.

Leave a Mark at *d*, and plant the Table at *e*, having firft meafured the Length of *d e*, 364 Links, and fet it on its proper Line from *d* to *e*, on which Line *d e*, lay the Index, and turning the Table about till you fee the Mark at *d*, there fcrew it faft, and having proved the Line *d e*, to be truly laid down in refpect of *c d*, turn the

Index

Index about on *e*, till you fee the Mark at *a*, and
draw the Line *e a*, which will cut through the
Point at *a*, alſo direct the Sights to the Angle *p*,
and draw ⊙ *p*, and to *q*, drawing ⊙ *q*, on the Paper.

Then meaſure with the Chain the Diſtances
⊙ *p*, and ⊙ *q*, ſetting thoſe Diſtances on their
proper Lines, and draw the Bounders *p o, q p*.

Now meaſure on the Line *e a*, and when you
come againſt the Bend in the Hedge at *r*, meaſure the Diſtance from the Chain Lines *e a*, to
that Bend 8 Links, which ſet on the Paper to *r*,
and draw the Bounder *q r*.

In the ſame manner meaſuring on the Line *e a*,
ſet off the Diſtances from the Chain Line to the
Angles in the Bounder ſ and *t*, and draw the
Bounders *r ſ, ſ t* and *t f*, which croſſes the Line *g*
f, near *f*, and when you have meaſured the Line
e a, on the Ground, you will find it the ſame
Length as that before drawn on the Paper.

If the Diſtance from the Station to the Bends
in the Hedge be great, 'tis the ſafeſt way to plant
the Center of the Table over the Station Line, as
at *w*, and laying the Index on the Line *e a*, direct
the Sights to *e*, or *a*, by turning the Table about,
and in that Poſition ſcrew it faſt; then direct the
Sights to the Bends in the Hedge as, to ſ, or *t*,
drawing Lines towards them, and ſetting off the
Diſtances in the ſame manner as is done from ⊙ *e*.

In order to examine the Lengths and Poſitions
of each Line before you proceed on the next, if
you don't think it convenient to leave Marks at
all the Stations round the Field; if you turn
about the Index on the Point, repreſenting the
preſent Station, till you ſee any one of the Marks
before laid down, and if the Edge of the Index
cuts that Point on the Paper, your Work is right;
ſo you might have left a Mark at the Firſt Station

e

a, and by that prove the Lengths and Positions of all the other Lines, as well as by the last but one you passed by.

But if you could not see the Mark at *a*, at all the other Stations, you may make use of any other Mark, as some Part of the Barn; or you may set up a Mark in some convenient Place, from whence you can see all the Angles.

But if the Mark you last us'd, is at too great a Distance from you, or lyes almost in a Strait Line with that you last laid down, then use some other Mark in its Stead, whose Position you have before found at one of the foregoing Stations.

Or instead of a Mark thus set up, you may use any remarkable Tree, Steeple, &c. that is not at too great a Distance from you, whether it be in the Land you are then Surveying or not.

How to Measure an inaccessible Distance

Lastly, In the same manner as you found the Position of the Barn in the middle of the Field, *Fig.* 11. you may measure any other inaccessible Distance; so if the Barn was so situated that you could not come nearer thereto, by reason of Water, or other Impediments, than the Line *b c*, yet you would know its Distance from *b*, or *c*.

The Instrument planted at *b*, and the Sights directed to a Mark at *c*, and also to the Barn, and the Lines *b z*, and *b c*, drawn on the Paper as before directed, and the Instrument removed from *b*, and planted at *c*, and the Sights directed to a Mark left at *b*, and also to the Barn, and the Line *c x*, also drawn on the Paper, cutting the other Line *b z*, then shall the Point of Intersection determine the Distance of the Barn from *b* or *c*, which you may find by measuring

from

from *b* or *c*, or any other Part of the Line *b x*, by the fame Scale with which you laid down the Line *b c*.

But 'tis convenient to make the Stations *b* and *c*, at fuch Diftance from one another, that the Angle at the Point of Interfection may not be too Acute, left you not be able to diftinguifh nicely the Point of Interfection.

If *Fig.* 11. was a Wood, fo that you could not meafure the Station Lines within, you may as well make them Outfide, for the Plott will be the fame, only the fmall Pieces of Ground between the Station Lines, and Bounders, are excluded by the Bounders from being any Part of the Plott. Particular Directions in this Cafe are needlefs; *fee the Figure.*

When you are about to meafure a Plott of Land, and in doubt whether it will lye on one Sheet of paper, you may place a Line or two croffe the Plott, as you walk about to choofe the moft convenient Stations (reckoning fo many Steps to a Chain, as you find by Experience carries you a Chain's Length) and thereby guefs what Scale to make ufe of.

Alfo, 'tis convenient to make a particular Remark at the Firft Station in each Field, that you may readily find it when you come round to it again, in order to clofe the Plott.

Alfo as foon as you have drawn the Plott of a Field, 'tis neceffary to write the Name fomewhere in the middle thereof.

There is another Way of Plotting a Field by the Plain Table (though fcarce fit for Practice) by meafuring one Line only, which in fhort is this:

Plant the Table at *a*, *Fig.* 12. and direct the Sights to the feveral Angles round the Field, keeping the Index clofe to the Point *a*, and by

the

the Edge thereof draw a Line towards every Angle, then place a Staff at *a*, and plant the Table at the other Station *b*, meafuring the Diftance between the two Stations, which Diftance fet on the Line *a b*, lay the Index on the Line *b a*, and turn the Table about till you fee the Mark at *a*, and fix there the Inftrument.

Then lay the Index to the Point *b*, and turn it about thereon, directing the Sights to the feveral Angles round the Field as before at *a*, and towards every one of them draw a Line, which will interfect the Lines before drawn at *a*, fo fhall the Points of Interfection determine the Place of all the Angles round the Field, and Lines drawn from Point to Point fhall give the Bounders of that Field.

But if you be not very exact and curious in drawing the Lines, and alfo, if the Stations are not contrived in fuch manner as may prevent the Lines Interfecting one another at very acute Angles you may commit grofs Miftakes.

And here it may be obferved in this as well as any other Cafe of the like Nature, that all Things that are to be determined by the Interfection of Right Lines are beft determined when thofe Lines Interfect each other nearest Right Angles.

Therefore when Triangles laid down by the Length of their Sides interfecting one another with acute Angles, the Point determining where thofe Lines interfect, is not fo well determined as it ought to be.

C H A P.

CHAP. II.

Shewing how to Survey any Piece of Land, by the Theodolite or Circumferentor, and to protract the same.

SECT. I.

T H E Plain Table is very ufeful for taking the Ground Plott of Buildings, and meafuring Gardens, or fmall Enclofures (where the fhortnefs of Lines, and Multiplicity of Angles would be apt to breed Confufion in protracting,) but by no means fit for furveying large Tracts of Land, becaufe the leaft Moifture, or Dampnefs in the Air, makes the Paper not only fink, but run up when dry'd again, and thereby the Lines drawn thereon are diforder'd, making the Content lefs than it fhould be; and in the leaft Rain or Mift, the Inftrument becomes altogether ufelefs; alfo, when the Plott proves larger than will lye on one Sheet of Paper, there muft more be pieced thereto with Glue or Pafte, which wetting only fome Parts of the Paper, is liable to the aforefaid Inconveniencies, neither can feveral Sheets of Paper be joyn'd together after the Plott is drawn thereon, fo as to meet exactly, and lye fo flat as it
ought

ought to do; and if to thefe Inconveniences be added the tedioufnefs of compleating the whole Plott in the Field, when a Surveyor has his Affiftants about him, that alone might be Objection enough to induce any Perfon to make ufe of fitter Inftruments.

Here follows the Defcription and Ufe of a New Theodolite, being the moft abfolute Inftrument yet invented for Surveying Land.

If we be not very exact and curious in mea- furing the Angles in the Field, the Plott on the Paper can never be truly laid down; and if the Inftrument with which we meafure thefe Angles be not well fram'd and divided, all the care we can take in making the Obfervations in the Field will be to little purpofe; therefore, I thought it might not be improper firft of all, to give a ge- neral Defcription of a new Theodolite, which hath met with a general Applaufe from all Ma- thematicians that have feen it, and far exceeds any other Inftrument that hath yet been in- vented.

For the Ball and Socket is fo contriv'd, that the whole Inftrument may be fet as truly Hori- zontal, as by the help of any Rack, and with lefs than a tenth Part of the Trouble and Time, and this in meafuring every Angle is abfolutely neceffary; for Inftance, fuppofe we were to meafure an Angle on the fide of an Hill, by one of the Theodolites as commonly made, and had fet the Inftrument as near a level as we could by the Eye, and then proceeded to make an Ob- fervation; if then the Limb be out of the Hori-

zon,

zon, fuppofe but two Degrees; (and it can fcarce
be fet nearer if fo near, for the Ground being on a
Declivity will deceive us) the Angle thus
meafured will be confiderably falfe; for the
Inftrument thus planted on the fide of the Hill,
let the Tellefcope be directed to the firft Object,
and that part of the Limb next your Eye fixed two
Degrees out of the Horizon, and then let the Tellefcope or Sights be turn'd round to the other Object,
and elevated thereto, fuppofe ten Degrees; then
the Index will cut on the Limb, the Number
expreffing the Angle, fuppofe ninety Degrees;
now this Angle of ninety Degrees is meafured
above twenty one Minutes falfe, and if the Lines
that form the Angle, happen to be long, this
error will be very confiderable in regard of the
true Content of that Piece of Land, and the Plott
can never be expected to clofe, if the Angles be
thus meafured; but in the ufe of this Theodolite,
this Inconvenience is remedied, the Inftrument
being fo eafily fet exactly Horizontal.

There is a Quadrant nearly, the Radius of the
Inftrument fo fix'd over the Center, as to move
exactly in a vertical Circle, within which is a
Spirit level, and over that the Telefcope fix'd
thereto, fo contriv'd that when the Bubble refts
in the middle of the Spirit-tube, the Horizontal
Hair in the Telefcope will cut an exact Level,
and by its Motion in a vertical Circle, whatever
Object this Hair cuts above or below, the true
Level, its Elevation or Depreffion will appear by
the faid Quadrant divided and grav'd for that
Purpofe; there is alfo in the Telefcope a vertical
Hair to be us'd in the meafuring Horizontal
Angles, fo that both the Horizontal and vertical
Angles are obferv'd at the fame time, which is
extream Ufeful in laying down the Plotts when
the

the Hynothenufal are to be reduc'd to horizontal Lines.

The Inftrument is well contriv'd for working with the Needle only; for as foon as the Inftrument is fet fteady on the Ground; the Needle will lye in the direction of the magnetick Meridian, and there be at reft; and then the Point in the Box mark'd with 360, may be brought to the north End of the Needle, and there fix'd without ftirring the Needle; alfo the Index and Telefcope may be mov'd round to any Object in the fame Manner; for the Head of the Staff is made of Brafs, and not liable to fhake as the wooden ones are, which contributes much to the true meafuring of an Angle, and the Index is mov'd round a conical Center, touching the Limb in three Places at 120 Degrees Diftance, and if by much wearing it fhould fhake, that is inftantly help'd by a Screw for that Purpofe; for if the Index grows loofe and fhakes, it will not cut the Minutes on the Limb to any Exactnefs.

The Pin, on which the Needle hangs, is made of temper'd Steel turn'd and polifh'd in the Lath, and may be taken out and put in at Pleafure, and is not fcrew'd to the Box, but remains fix'd always in the fame Pofition, whilft the Box, Index, and Telefcope are mov'd round it, and the Telefcope fix'd to an Object without ftirring the Needle, fo that an Obfervation may be made both by the Limb and Needle at the fame Inftant.

And when the Telefcope is directed to an Object, the whole Inftrument is fix'd there in fo firm a Manner, with fmall Power, that the Motion of the Index when the Telefcope is directed to the next, fhall not move the Limb from the Pofition in which it was firft fix'd, which in other Theodolites is very difficult to be done.

The

The whole Inftrument is made very portable, and the Ufes thereof plain and fimple; one Telefcope being apply'd to all the Operations, neither is any thing to be added or taken from it when we make ufe thereof.

Being thus provided with a good Theodolite; we now proceed to fhew the Manner of ufing it in meafuring Land.

The moft material Things to be done in the Field are two; Firft, to meafure the Length of the Lines; and Secondly, the Quantity of the Angles.

The manner of meafuring Lines in the Field is already laid down; we now proceed to the Angles.

SECT. II.

The Defcription of an Angle.

An Angle is the metting of the two Lines in a Point; provided the two Lines fo meeting don't make one ftrait Line, as the Lines *a, o,* and *x, o, Fig.* 12. meeting together in the Point *o,* form the Angle *a, o, x.*

Angles are meafured by the Arch of a Circle defcrib'd from the angular Point as a Center; fo the Angle *a, o, x,* is meafur'd by the Arch of the Circle, *Fig.* 13. defcrib'd from the angular Point *o:* The Arch of this Circle that meafures the Angle *a, o, x,* being intercepted between the Lines *a, o,* and *x o.*

An Angle is faid to be equal to, greater or lefs than another, according as the Arch which meafures it contains as many more, or fewer of the equal Parts into which the Circle is fuppos'd to be divided.

The Circle is divided into 360 Parts or Degrees,

grees, and each Degree into 60 other Parts call'd Minutes; or suppos'd to be so divided, so that any Portion of the Circumference is expres'd by the Number of Degrees and Minutes it contains.

'Tis no matter whether the Circle be great or small, for each is suppos'd to contain 360 Degrees (except that the Minutes are better estimated on a great Circle than a small.)

The Line *o h*, is the Radius of the Circle, *Fig.* 13. *z*, *d*, is the Diameter, and passes through the Center *o*, dividing the Circle into two equal Parts; the Line *z e*, is the Chord of 60 Degrees joining the Extremities of the Arch *z e*.

Z o h, is a Right angled Triangle, *z o* the Base, *o h* the Perpendicular, and *h z* the Hypothenusal. Then the Angle *z o h*, contains 90 Degrees, which is the Measure of a right Angle.

A o x, is an acute Angle, containing less than 90 Degrees.

A o y, is an Obtuse Angle, containing more than 90 Degrees.

'Tis no matter of what Length the Lines of an Angle be; 'tis their meeting one another in a Point that forms the Angle, for the Angle *z o h*, contains 90 Degrees, and *a o c*, contains 90 Degrees also.

Let the Circle *Fig.* 13. represent the Limb of the Theodolite, and let the Index be brought to the beginning of the Numbers at *z*, then the Telescope which moves therewith, will be set in the Direction of *o a*: Now move the Telescope till it be in the Direction of *o x*, so shall the Index on the Limb shew the Number of Degrees of the Angle *a o x*.

Let *x o*, and *o y*, represent two Station Lines in the Field, then the Center of the Theodolite being planted over the angular Point *o*, let the

Telescope

Telescope be directted to *x*, (the Index being brought to the beginning of the Numbers on the Limb at *o*) and the Instrument fix'd there: Now move the Telescope till you see *y*, so shall the Index shew the Quality of the Angle *x*, *o*, *y*, on the Limb, 119 Degrees.

In working with the Theodolite we only measure the Length of the Lines, and the Quantity of the Angles in the Field, which we note in a Field-Book for that Purpose; and by these Directions we may draw the exact Plan of the Land, which is call'd Protracting.

Provide a Field-Book rul'd with three Columns, in the middle Column insert the Quantity of the Angles, and the Length of the Station Lines; in the outer Columns the Offsets from the Station Lines to the Bounders, and on each Side note the Remarks which you meet with in the Survey.

SECT. III.

Shewing the Uses of the Theodolite in measuring the Angles round any Field or other Piece of Land.

The Theodolite before mention'd is numbred on the Limb, from the Left to the Right, and the Box from the Right to the Left; and therefore the Circuit is most conveniently made (according to the common Phrase) *contra solem.* i.e. with the Fences on the right Hand.

Let *Fig.* 14. represent a Field of which a Plott is desired: First, chose some convenient Place therein, to being the Work as at ☉, near the Corner of the Field *a*, and set up a Staff with a Mark thereon, as a Piece of Paper, or a white Rag, so that you may plainly see it at the next

Station;

Station; then lay the Chain in a ſtrait Line from *a* towards *b*, having firſt ſet up a Staff at *b*.

The Chain lying in this Direction on the Ground, meaſure with the Offset Staff from ⊙, the Diſtance to the Hedge 56 Links, and enter in the middle Column of the Field Book *o*, and in the right Hand Column 56, denoting an Offsett laid off from ⊙, 56 Links to the right Hand of the Station Line.

Meaſure forwards on the Station Line *a b*, till you come againſt the next Bend in the Hedge, then let the Chain lye on the Ground in a right Line between the Stations *a*, and *b*, and with the Offset Staff, meaſure the Diſtance from the Chain to that Bend in the Hedge 140 Links.

Take notice at what Length of the Station Line each Offset is laid off; as here at the Length of 540 Links on the Station Line, I lay of an Offset of 140 Links, therefore in the middle Column of the Field-Book that repreſents the Station Line enter 540, and againſt it in the Column on the right Hand, write 140, denoting that at the Length of 5 Chains 40 Links in the Station Line, you laid of an Offset to the right 140 Links.

Alſo take Notice that theſe Offsets are to be meaſured from the Station Line to the Hedge or Bounder, in ſuch Manner that the Line repreſenting the Offset may ſtand at right Angles with the Chain or Station Line, as the Line *o c*, with *z d. Fig.* 13.

Here I would adviſe a young Beginner in this Art, not only to enter theſe Obſervations in the Field-Book, but alſo on a Piece of waſte Paper, to draw firſt a Line that ſhall repreſent the preſent Station Line, and then upon that, to ſet of the Offsets as he meaſures them in the Field,

drawing

drawing the Bounders as he goes along, not regarding the Length of any Line, a resemblance of
each Line being sufficient; because the true
Lengths are entered in the Field-Book; for
laying this Sketch before him, when he protracts
his Work, he will find it an useful Instruction, in
drawing his Angles inward or outward, and
laying the Corners of the Field in their true
Position, with other little Difficulties to be met
with in Practice; but when he is accustom'd to a
right Method of keeping the Field-Book this
Trouble will be spar'd.

I proceed to measure on the Station Line to
the next Bend in the Hedge, and there lay of an
Offset at 8 Chain 26 Links, therefore against 826
in the middle Column of the Field-Book; write
in the next Column on the Right; 36 Links the
Length of the Offset.

The Hedge continuing streight to ☉ 2,
measure up thereto, and enter the Length in the
Field Book 1120, and measure the Distance to the
Hedge 36 Links, which enter in the Column of
Offsets against 1120, and draw a Line cross the
Field-Book.

Now remove the Staff from *b*, and cause it to
be set up in a convenient Place, as at *c*, then plant
the Center of the Theodolite exactly over the
Hole, in which the Staff stood at *b*, by the help of
the String and Plummet, as directed in the Use of
the plain Table, making the Staves of the
Instrument to stand firm on the Ground, then
bring the Index to 360 on the Limb, and turn the
whole Instrument about till the Hair in the
Sights cuts the Staff at *a*, and there Screw it fast,
that the Motion of the Index may not cause it to
stirr from this Position; then turn the Index
about till the Hair in the Telescope cuts the Staff

at

at *c*, so shall the Index shew the Quantity of that Angle *a b c*, on the Limb, *viz.* 102 Degrees 20 Minutes, which note in the Field-Book for the Quantity of that Angle.

Now for certainty you have measured this Angle Right, you may turn the Telescope back to the Staff at *a*, and if the Hair cuts it you are right, otherwise not.

Having measured this Angle, let the Staff be brought from *a*, and place it in the Hole, over which the Center of the Instrument was posited at *b*, but leave some Remark at *a*, that you may find it again when you come round the Field to close the Plott: and lay the Chain from *b*, towards *c*, and at ☉, measure the Offset to the Hedge 20 Links, at 236, in the Station Line, I lay of the Offset 36, at 428 in the Station Line, the Offset is 92, and at 796 the End of the Line, the Offset is 30, to the Corner, therefore against 30 in the Column of Offsets, write Corner, denoting that Offset laid of at Right Angles from the Station Line, reach'd the Corner of the Hedge.

Place the Instrument at *c*, and as before directed, measure the Angle *b c d* 110 Degrees 40 Minutes, which note in the Field Book for the Quantity of the Angle at *c*.

When you have measured the Angles, and made the necessary Observations at each Station draw a Line cross the Field Book as you will see in the Form thereof, also take notice that the Minutes are estimated by the help of *Nonus's* Invention which can't be so well describ'd as by the Sight of the Instrument, only this may be said, that we can thereby estimate the Quantity of an Angle to 2 or 3 Minutes, which is as exact as they can be laid down on Paper by the Protractor.

Bring

Bring the Staff from *b*, and set it as upright as you can at *c*, and send another forwards to *d*, then measure on the Line *c d*, and lay of the Offset to the Corner at 434, and against that Offset write Corner in the Field-Book, and measure up to *d*, entering the Length 468 in the Field-Book.

Plant the Instrument at *d*, and bring the Index to 260 on the Limb, and turn it about till the Hair in the Telescope cuts the Staff at *c*, and there fix the Instrument, and then direct the Telescope to *e*, and note the Quantity of the Angle at *d*, which the Index cuts on the Limb, *viz.* 230 degrees 50 Minutes, which note in the Field-Book for the Quantity of that Angle.

But no Angle is greater than 180 Degrees, therefore if you would know the true Quantity of this Angle, subtract 230 Degrees, 50 Minutes, from 360 Degrees, the Remainder is 129 Degrees 10 Minutes, the true Quantity of that Angle.

Note, When you meet with an outward Angle, remember to Mark it in the Field-Book with $>$, or some such Mark, as a Direction when you come to protract this Angle; to draw it outwards as it is in the Field.

In the same Manner deal with the rest of the Lines and Angles round the Field, till you come to Station *a*, but there is no necessity to measure the last Angle, or the two last Lines, unless it be to prove the Truth of the Work, which indeed is convenient.

When the Instrument was planted at *f*, and had measur'd the Quantity of that Angle, the Instrument remaining in the same Position, if you direct the Telescope to the Tree in the Middle of the Field, and note the Degrees, *&c.* which the

Index

Index cuts on the Limb, and the fame at *g*, and note thefe Degrees, *&c.* in the Field-Book, in the Column of Remarks, you may protract the true Situation of the Tree in refpect of any other Part of the Field.

See the form of thefe Obfervations as noted in the Field-Book.

The Field-Book.

Remarks	Offsets	Station Lines	Offsets	Remarks
		$a \odot 1$		
		- - - - -		
		0	56	
		540	140	
		826	36	
		1120	36	
		$b \odot 2$		
		o ′		
Angle		102.20		
		- - - - -		
		0	20	
		236	36	
		428	92	
		796	30	Corner
		$c \odot 3$		
		o ′		
Angle		110.40		
		- - - - -		
		434	30	Corner
		468		
		$d \odot 4$		
		o ′		
Angle >		230.50		
		- - - - -		
		420	30	

The

The Field-Book Continued.

Remarks	Offsets	Station Lines	Offsets	Remarks
		e ☉ 5		
		∘		
	Angle	79.00		
		- - - - -		
		∘	40	
		134	36	
		296	33	
		588	100	
		820	12	
A Tree bears		*f* ☉ 6		
from ☉ 6		∘ ′		
∘ ′	*Angle*	84.30		
38.30		- - - - -		
		40	120	*Corner*
		200	24	
		706	16	*Corner*
Tree bears		*g* ☉ 7		
from ☉ 7		∘		
∘ ′	*Angle* >	233		
57.30		- - - - -		
		380	80	
		648	40	

SECT. IV

The next Thing to be done, is to protract the Obfervations made in the Field, *Fig.* 14. fo that
the

the feveral Lines and Angles therein may be laid
down on Paper of the fame Length, and in the
fame Direction as in the Field it felf. In proportion as the Scale we make ufe of is to the Chain.

The Protractor generally ufed, is a Semicircle,
though a whole Circle is better. Numbred and
divided in the fame manner as the Limb of the
Theodolite, which it fhould always be; the
Protractor being efteem'd an Epitome of
Inftruments.

But becaufe the Degrees on the Protractor,
are fo much fmaller than thofe on the Limb of
the Theodolite, they can't be well eftimated
nearer than 10 Minutes; yet if any one will be
curious, he may lay down the Angles on the Paper, to a Minute or two as exactly as they can be
obferved in the Field.

Mr. *Ward's* Protractor being commonly ufed
for this Purpofe, is made with an Index to move
on the Center of the Semicircle, which Index is
divided into 2 Parts, fo fram'd, that each may be
the Diagonal of one Degree; fo that if the
diftance at the extream Ends be 10 Degrees, that
next the Limb muft be 8 Degrees, the Space
between the two Limits in each Diagonal being
divided into 60 Parts or Minutes; but thefe
Divifions will be very unequal, being thofe of the
Tangent Line, which fall near Infinite.

Mr. *Siffon* hath removed this Inconveniency,
by making each Edge the Arch of a great Circle
paffing through the Center of the Protractor; the
Space between the firft and laft Divifions being
two Degrees thereof, and is divided into 60 equal
Parts or Minutes.

The Reafon depends on the 27 *Prop.* of the 3*d*,
of *Euclid*, *viz.* That the Angle at the Center of a
Circle is double to that at the Circumference.

The

The fame Perfon hath contrived another Protractor, to lay down Minutes without any Index at all, and therefore preferable to both the former, becaufe 'tis exceeding difficult to make the Index move exactly round the Center, and if it fhakes the leaft that can be, the Inftrument is ufelefs; but this laft is made of one Piece of Brafs, and may be us'd as a common one, without regarding the Minutes when Expedition is requifite.

It may be thought here are too many Words fpent about defcribing thefe Protractors as well as the Theodolite before mentioned; but if any one pleafes to confider, that if we be not very exact in meafuring the Angles in the Field, and laying them down in the fame manner on Paper, (which is impoffible to be done without good Inftruments) we fhall commit very great Miftakes; for Inftance, If I miftake half a Degree in the meafuring of an Angle, one whofe Sides is 20 Chains, the Area or Content of that Piece of Ground fo left out or added to the Plott, by drawing this Line is a falfe Pofition, will be above 23 Poles; and this Error communicated to the following Work, will be very confiderable in the whole.

SECT. V.

The Manner of Protracting the aforegoing Obfervations.

AS the Lines are meafured in the Field by the Chain, and the Angles by the Limb of the Theodolite, fo the Lines are laid down on the Paper by the Scale, and the Angles by the Protractor.

Provide a Skin of Parchment, if the Plott is defired to be on Parchment, according to the
largenefs

largenefs of the Work you are about to lay down; or if on Paper, let it be large enough to hold all your Work; the ftrong Cartridge Paper for this Purpofe is accounted beft by fome Surveyors.

Having confidered which way the Plott will extend, draw an obfcure Line on the Paper to reprefent the firft Station Line, and mark the End thereof with ⊙ *a*, fo fhall that Point reprefent the firft Station in the Field, and clofe to this obfcure Line, lay the edge of your plotting Scale, the beginning of the Numbers coinciding with ⊙ *a*, and encreafing towards the next Station; then lay the Field-Book open before you, and becaufe the offsets in the firft Length are taken at the Diftances o, 540, 826, 1220; therefore againft thefe Numbers on the Scale, make Marks in the obfcure Line, clofe to the edge of the Scale.

This done, turn the Scale Perpendicular to the obfcure Line, fo that the feveral offsets may ftand thereon at right Angles as aforefaid, and apply it fucceffively to thefe feveral Points, and there Prick off the Length of the feveral offsets on the fame Side of the obfcure Line as noted in the Field-Book; fo at ⊙ I prick off 56 at 500, the next Length I prick off 140, at the next Point, which is at the Length 826, I prick off 36, and at 1120 the End of the Line, I prick off 36.

Now if the Lines are drawn from Point to Point, they fhall reprefent the Bounders of this Side of the Field; and becaufe the Hedges, efpecially in old Enclofures, are generally in the Form of a curve rather than ftrait Lines, therefore if you draw the Bounders from Point to Point with a Quill-Pen with your Hand only, they will be more naturally exprefs'd, then if you lay a ftrait Ruler from Point to Point, (except the Diftances are very long, or you take a multitude of Offsets;)

Offsets;) and to be exact, 'tis fometimes
neceffary to exprefs the Nature of thefe little
Irregularities in the Fences, by a Sketch on one
Side of the Field-Book; but if you will be very
curious, you may have an Inftrument in the
Form of a Steel-Bow, which by the help of
Screws may be drawn in any curve Form, and by
this the Bounders may be readily drawn.

The Length of the firft Station Line being
1120, mark that Diftance from ⊙ *a*, with ⊙ *b*, and
let the obfcure Line be produc'd, each way as
long as the Radius of the Protractor.

Lay the Center of the Protractor to the Point
⊙ *b*, and turn it about thereon, till the Diameter
lyes on the Line ⊙ *a*, ⊙ *b*, the beginning of the
Numbers on the Protractor being laid towards
⊙ *a*, contrary to the Theodolite in the Field.

Hold the Protractor clofe down to the Paper
in this Pofition, and becaufe the Angle at *b*, is 102
Degrees 20 Minutes, therefore with a Protracting
Pin or Needle, make a Mark againft 102 Digrees
20 Minutes, clofe to the Limb of the Protractor,
through which Mark from *b*, draw the obfcure
Line *b c*.

So is the Station *b c*, laid down in the fame
Direction as in the Field, and the Angle *a b c*, the
fame.

Lay the Plotting Scale to the obfcure Line *b c*,
the beginning of the Numbers coinciding with
the prefent Station, and the Numbers encreafing
towards the next, then clofe to the edge thereof,
againft 0, 236, 428, 796, the Lengths where the
Offsets were taken, make Marks with the
Protracting Pin, and turn the Scale perpendicular
to the obfcure Line, and Prick off the feveral
Offsets, 20, 36, 92, 30.

And now if Lines are continued from the
Fences

Fences before drawn to thefe Offsets, they fhall reprefent the Bounders on this Side of the Field.

The Offset at the End of the fecond Station Line, at *c*, reaches into the Corner, but thofe at *b* muft be continued till they meet one another, and this might be expreffed in the Field-Book or Sketch, that you may not miftake the Corner of the Field.

Lay the Center of the Protractor to *c* the Diameter, held clofe to the Line *b c*, and againft 110 Degrees 40 Minutes on the Limb of the Protractor, make a Mark, through which draw the Line *c d*.

At the Length 434, in this Line lay of the Offset 30 Links, to which continue the Bounders before drawn, fo is this Side of the Field finifhed.

Note, the next Angle at *d*, being noted in the Field Book, 230 Digrees 50 Minutes, you muft either fubtract 2°30 5'0, from 3°60:'00 the Remainder is, 129 Degrees 10 Minutes for the true Quantity of that Angle; and becaufe 'tis marked External, it muft be plotted outward, and the beginning of the Numbers on the Protractor muft be laid the contrary way, *viz.* towards the next Station.

Or if the Protractor be numbred to 360, on a Circle concentric to the outward Circle, and the Numbers on both encreafe the fame way, (as the Limb of the Theodolite) then the Angle 230 Degrees 50 Minutes, may be pricked off from the inner Circle.

But if you ufe a circular Protractor, it may be laid always one way, *viz.* the beginning of the Numbers towards the laft Station, contrary to the Theodolite in the Field, and this in my Opinion is the beft Way.

But if you ufe a femicircular Protractor, obferve

ferve to lay the Diameter on that Line which brought you to the prefent Station, and to lay the beginning of the Degrees of the Protractor towards the laft Station when the Angle is lefs than 180 Degrees, but the contrary way when the Angle is more.

So at *d*, lay the Diameter of the Protractor on the Line *c d*, the beginning of the Numbers being laid the contrary way to *c*, and againft 230 Degrees 50 Minutes on the inner Circle of the Protractor, make a Mark, through which draw the Line *d e*.

In the fame manner lay down the Angle at *e*, and draw the Line *e f*, continuing the Bounders as before directed.

When you have marked the Angle at *f*, let the Protractor lye in the fame Pofition, and make a Mark againft 38 Degrees 30 Minutes, as noted in the Field-Book for the bearing of the Tree from that Station, and through that Mark draw an obfcure line from *f*.

Do the fame at *g*, continuing the obfcure Line from thence till it croffes that drawn from *f*, fo fhall the Interfection of thefe two Lines determine the Situation of the Tree in the Middle of the Field.

In the fame manner may any other inacceff-ible Diftance be meafured by the Theodolite.

When you have Marked the Angle at *g*, and drawn the Line *g a*, it will cut through the Point at *a*, and the Length of the Line *g a*, will be the fame as that noted in the Field-Book, and the Angles at *g*, and *a*, (if you had meafured it) the fame which proves the Plott to be truly laid down.

SECT.

SECT. VI.

IT may not be improper to take notice in this Place, of the Method propofed by fome Authors, as a Proof that the feveral Angles in a Field are truly meafured, by collecting the Quantities of all the Angles into one Sum, and then to multiply 180, by a Number lefs by two than the Number of Angles in the Field; and if the Product of this Multiplication be equal to the total Sum of the Angles, the Work is concluded to be right.

But thefe two Numbers may agree, and yet a Miftake may be committed in meafuring the Angles; as for Inftance:

Let the Number of Angles in the Field be 7, and the Quantities collected into one Sum be 900; then multiplying 180, by a Number lefs by two than the Number of the Angles, *viz.* 5, the Product is 900, equal to the fum of the Angles.

Let the true Quantity of the firft Angle be 160 Degrees, and the true Quantity of the Second 190 Degrees; thefe two Numbers when added together make 350; but fuppofe you had made a Miftake in eftimating the Degrees on the Inftrument, or noting them in the Field Book, and for the Firft Angle had noted 190 Degrees, and for the Second 160 Degrees, their Sum will ftill be 350; fo that by this Method you will not difcover your Error; but hereafter will be inferted a Method whereby an Error may be corrected at every Station in the Field before we leave it by the help of the Needle and Limb together, but firft I proceed to fhew the Ufe of the Needle only in furveying Land.

SECT.

SECT. VII.

Of the Circumferentor.

THE Circumferentor is an Inftrument ufed to meafure Angles in the Field; it confifts of a Box and Needle, fcrewed to the Index with plain Sights thereon, or inftead of the plain Sights with a Telefcope mounted over the Box, that may be either elevated or depreffed to an Object as there fhall be occafion; the Index is mov'd by a Ball and Socket, and fupported by a three legged Staff.

In Surveying Harbours, Seacoafts, Counties, or large Commons, where the Lines are very long, or thick over grown Woods, where we may be forced to make a multitude of Angles, and the Sight of the two Lines conftructing the Angle, may be hindered by the Brufh or Underwood; in thefe Cafes the Angles may be meafured fufficiently exact by the Needle only, (though better, and as quick by the Theodolite, as will be fhew'd hereafter) yet in furveying Lordfhips, Enclofures, or plain Pafture Land, (a fmall piece of which got or loft is of confiderable Value, and each particular Field ought to clofe exactly) the Angles are without doubt more furely meafured by the Limb of the Theodolite, becaufe the Degrees in the Box can't be fo nearly eftimated, and the Needle is liable to be drawn afide by fome hidden magnetick Power.

The Pofition or bearing of a Line obferv'd by the Needle, is expreffed by fuch a Number of Degrees and Minutes as it is diftant from, or Quantity of the Angle, which that Line makes with the Meridian.

And if a Perfon wholly unacquainted with the
ufe

ufe of this Inftrument, will take the Pains to try this following Method, it may be an help to conceive the manner of ufing it in the Field.

Upon a Sheet of Paper let there be drawn right Lines parallel one to another at any Diftance, and upon a Table let there be fixed a Pin with the Point upwards, let the Pin fo fixed, be run through one of the Lines in the Paper, and upon the Point of the Pin, let there be put a Magnetick Needle, let it traverfe about till it refts of it felf; then turn the Paper about on the Table till the Needle hangs directly over the Line, in which the Pin is placed, which is difcovered by fixing the Eye over its center; then with fealing Wax faften the Paper to the Table by the four Corners; fo may the Paper be fuppofed to reprefent the Surface of the Earth, and the Lines the magnetick Meridian (which mark at the top with North, and at the bottom with South.)

For if the Pin be removed into any other of the Lines, and the Needle be made to traverfe thereon, it will, when at reft, hang directly over the Line in which the Pin is placed, if it be drawn parallel to the firft Line, over which the Needle hung when the Paper was fixed.

The Needle then points always to or lyes in the direction of the Meridian, by virtue of the magnetick Power; fo if I had faftened to the Table a Sheet of blank Paper, and had laid a Ruler in the fame direction with the Needle when at reft, and had drawn a meridian Line, and removed the Needle to another part of the blank Paper, and drawn another fuch a Line by the direction of the Needle, that would have been a Parallel.

When we take an Angle in the Field by the
Needle,

Needle, the meridian Line is always one fide of the Angle, and the Hedge Wall or Fence along which the Telescope is directed, is the other fide of the Angle, and they are fuppofed to meet at the Center of the Inftrument.

But with the Theodolite, the Angle is formed by the meeting of the two Lines of Fences themfelves.

Set one Foot of a Pair of Compaffes in fome one of the Meridians on the Paper, and defcribe a Circle, then the Line is its Diameter: Divide this Circle into 360 Degrees, which is eafily done by the Protractor, and let the Numbers being at N. or North, and encreafe to the left, towards E or Eaft.

Then this Circle reprefents the Box of the Inftrument in the Field, and the Line N. S. reprefents the Needle.

From the Center of the Circle, draw a long Line any way at a venture, and image this Line reprefents an Hedge or Station-Line in the Field, and to find its bearing or Angle that it makes with the meridian, look what Degree, *&c.* it cuts on the Circle, for that is the Quantity of the Angle or Number, expreffing its bearing, counted from the beginning of the Numbers.

So the Needle ufed in the Field points out the magnetick Meridian, and the Divifions in the Box mov'd under it meafure the Angle, that any Line in the Field makes with that Meridian.

The Box of the Circumferentor is commonly numbred from the right to the left; the Numbers beginning at N or North, which is mark'd alfo with a Flower de luce, and encreafe towards E or Eaft, and the direction is to be taken from the North End of the Needle.

Let it be required to obferve the bearing of the feveral Station-Lines that encompafs the Wood, *Fig.* 15.

S E C T.

SECT. VIII.

The Use of the Circumferentor in Surveying Land.

Irst plant the Circumferentor at some convenient Station as at *a*; the Flower de luce in
the Box being from you, direct the Sights to a
Mark at the next Station *b*, and mark the Division which the North end of the Needle points
to in the Box when at rest, which is 260 Degrees
30 Minutes; therefore note this Number 260
Degrees 30 Minutes in the Field-Book, for the
bearing of the Line *a b*.

Observing former Directions for removing
the Instrument from one Station to another, and
measuring the Station-Lines and Offsets from
thence to the Bounders as you pass along the
Station-Lines, let the Instrument be removed
from *a*, and planted at *b*, the next Station; then
keeping the Flower de luce in the Box from you;
turn the Instrument about till the Hair in the
Sights cuts a Mark at the next Station *c*; then
will the North end of the Needle point to 292
Degrees 12 Minutes, which note in the Field-
Book for the bearing of the Line *b c*.

The Instrument planted at *c*, and the Sights
directed to *d*, the bearing of the Line *c d* will be
331 Degrees 45 Minutes.

In the same manner proceed to take the bearing of the other Lines round the Wood, observing this general Law.

To keep the Flower de luce in the Box from
you, and to take the bearing of each Line from
the North end of the Needle.

The Numbers in the Card of some of these
Circumferentors are made to encrease towards
the right, but that before mentioned is best; for
when

when you turn your Inftrument to the Eaftward, the Needle will hang over the Weftward Divifion on the contrary Side.

Inftead of planting the Circumferentor at every Station in the Field, the bearings of the feveral Lines may be taken if it be planted only at every other Station.

So if the Inftrument had been planted at *b*, and the Flower de luce in the Box kept toward you when you look back to the Station *a*, and from you when you look forwards to the Station *c*, the Bearings of the Lines *a b*, and *b c*, would be the fame as before obferved; alfo the Bearings of the Lines *c d*, and *d e*, might be obferved at *d*, and *e f*, and *f a*, and *f*; fo that inftead of planting the Inftrument 6 times, you need in this cafe plant it but 3 times, which faves fome Labour.

But fince you muft go along every Station Line, to meafure it or fee it meafured, the trouble of fetting down the Inftrument is not very great, and then alfo you may examine the Bearing of each Line as you go along; and if you fufpect an Error in the Work by the Needles being acted on by fome hidden magentick Power, or from your own Miftake, in obferving the Degrees that the Needle points to, you may correct fuch Error at the next Station before you proceed.

As when the Inftrument was planted at *a*, and the Sights directed to *b*, the Flower de luce from you, the North end of the Needle pointed to 260 Degrees 30 Minutes; now being come to *b*, direct the Sights back to a Mark at *a*, keeping the Flower-de-luce towards you: So fhall the North end of the Needle Point to 260 Degrees, 30 Minutes, as before at *a*, and then you may be fure the bearing of the Line *a b*, is truly obferved.

But if the Needle doth not point to the fame
number

number of Degrees, &c. there hath been some
Error in that Observation, which must be corrected before you proceed.

If you have a suspicion that the Needle doth
not play well, when the Instrument is planted at
any Station, as at *a*, direct the Sights to the Mark
at *b*, and note the Degrees, &c. pointed at by the
Needle in a piece of waste Paper; then with a
clean Knife, Key, or any bit of polish'd Steel, that
hath touched a Loadstone, move the Needle by
applying it to the Box, and examine when it
hath settled again what Degrees it then Points
at, the Sights being still directed to the
preceding Mark at *b*; and if the Degrees are the
same, they may be entred in the Field-Book, but
if not, the Cap and Pin must be cleaned with
some brown Paper and a little Putty, and thereby
freed from such Dust or Dampness that hath
gotten to it; if after all the Needle does not play
freely, place in the Box another Pin, or use
another Needle, or do both, and these Necess-
aries a Surveyor ought to have in his Pocket
while he is in the Field.

If you would measure the Quantity of any
Angle by the Needle, place the Instrument at the
angular Point, and take the Bearing of the two
Lines constructing that Angle, and subtracting
the lesser out of the greater, the Remainder is the
Quantity of that Angle, if less than 180 Degrees,
but if the Remainder is greater than 180 Degrees,
subtract it out of 360 Degrees, and that last
Remainder is the Angle.

The manner of entering the Offsets in the
Field-Book, is before shewn in the use of the
Theodolite; it will be sufficient in this place, to
insert the Bearing of each Line or Quantity of
the Angle, which each makes with the Meridian,
                                                together

together with their Lengths, in order to protract
or lay them down on the Paper Plott of the same
Length and in the same Direction as in the Field.
Vide Fig. 15.

SECT. IX.

*The manner of Protracting the aforegoing
Observations made by the Circumferentor.*

Lines	Bearings		Links	
a b,	260	30	1242	*First*, draw Lines
b c,	292	12	1012	parallel to one another
c d,	331	45	1050	quite through the
d e,	59	00	1428	designed Draught, at
e f,	112	15	645	Distances not exceed-
f a,	151	30	1806	ing the Breadth of the

diametrical Part of
your Protractor, as in *Fig.* 15, and mark them
with N, and S, for North and South; then
considering which way the Plott will extend,
assign a Point in some one of the parallel Lines, to
represent the first Station in the Field, as at *a*, to
which Point lay the Center of the Protractor, and
by the help of the Divisions continued beyond
the Ends of the Diameter of the Protractor, lay
the Diameter upon, or parallel to those North
and South Lines; the beginning of the Numbers
on the Protractor towards that part of the Line
mark'd with N, or Northwards, when the
Degrees are fewer than 180, but Southwards
when more; the Protractor thus placed, look in
the Field-Book for the Bearing of the first Line
a b, which is 260 Degrees 30 Minutes; therefore
with the beginning of the Numbers on the

Protractor towards ſ, cloſe to the Limb againſt 260 Degrees 30 Minutes make a Mark, and through that Mark from the aſſigned Point at *a*, draw a Line *a b*, on which Line ſet 12 Chains 42 Links, as noted in the Field-Book.

So will the Line *a b*, on the Paper, have a Bearing like to that, which you obſerved the Line *a b* to have in the Field, in reſpect of the Meridian, but the Protractor to lay down theſe Obſervations muſt be numbred contrary to the Box of the Circumferentor; and if it be a Semicircle it muſt be numbred, firſt to 180, and then on the inner Circle whoſe Numbers muſt encreaſe the ſame way as the outer Circle to 360, and the Bearings greater than 180, are pricked off from this inner Circle, and the beginning of the Numbers muſt be laid Northward or Southward as the Degrees of Bearing are more or leſs than 180; but if your Protractor be a whole Circle, the beginning of the Numbers may be kept always one way, as the Numbers of the Circumferentor were in the Field, (the Protractor being an Epitome of the Inſtrument you make uſe of in the Field) but the Diameter muſt be always laid upon a Parallel to the meridian Lines, and may be mark'd with N S at the Ends as a Direction to keep it in its true Poſition.

Having made the Line *a b* of its true Length and Poſition, the next thing to be done is to lay of the Offsets therefrom, which gives the Bounders of that ſide of the Wood, *Fig.* 15.

Lay the Center of the Protractor to the Point *b*, and becauſe the Bearing of the Line *b c*, is more than 180, lay the beginning of the Numbers of the Semicircular Protractor towards S, and againſt 292 Degrees 12 Minutes, make a Mark, through which Mark from *b*, draw the Line *b c*,

ſetting

setting of the Offsets therefrom, and draw the Bounders of that side of the Wood.

In the same manner lay down the other Lines *c d*, *d e*, *e f*, and *f a*; so will the Line *f a*, cut through the Point *a*, and be of the same Length on the Plot as that measured in the Field, if the Observations be truly made.

Then if you drew the Station-Lines, and Offsets with a black-lead Pencil, and the Bounders with Ink; you may with a Piece of Bread rub off those Lines, so shall the true Bounders of the Wood only remain, which gives the exact Figure thereof.

SECT. X.

The manner of casting up the small irregular Pieces of Ground, which lye between the Station Lines and Hedges.

IT very rarely happens that the sides of a Field are all strait Lines, and therefore any Method for measuring them from one or more Stations in the Middle, can seldom be put in Practice; the best way being to go round, and measure the several Angles from Stations near the Bounders, but at such a Distance from thence that we may see clearly from one Station to another, and have plain Ground to measure the Distances, free from the Incumbrance of brushwood, Trees, &c. so shall the greatest Quantity of the Land be included between the regular Station-Lines, which is cast up as before directed by dividing the same into the largest Trapezias and Triangles that may be, and measuring the Bases and Perpendiculars by the same Scale that the Plott was laid down by.

But in order to cast up the small irregular

Pieces

Pieces comprehended between the Station-Lines and Bounder; if you reduce them into Triangles, &c. as they will be a great many in Number, fo you will very much err in laying of them down firft, and taking them off afterwards, efpecially if the Scale you protract by, be very fmall, where 10 or 12 Links of a Chain is hardly to be eftimated though the Scale be well divided, and the Points of the Compaffes very fine: For the removal of this Inconvenience, I fhall here fhew a way whereby you may caft up thefe fmall Quantities, let the Scale be never fo fmall, as exactly as any of the greater parts of the Field.

Suppofe the fmall irregular Pieces between the Station-Lines and Bounders, *Fig.* 14, were to be caft up.

Firft lay the Field-Book before you, where you will find the Length of the firft Offset (as meafured in the Field with your Offset Staff) from ⊙ 1 at *a*, to be 56 Links, and the fecond at 540, in the Chain-Line 140, forming the fmall Trapezia, *a, Vide Fig.* 16.

Now if you add the Offset 56 to the next 140, the Sum is 196, the half of which is 98, the equated Breadth; multiply the Length 540 by 98, the Product is 52920, the content of the Trapezia, *a*, in fquare Links.

Add 140 to 36, the fum is 176, the half Sum 88, fubtract 540 from 826, the Remainder is 286, the Length of the Trapezia, *b*; therefore multiply 286 by 88, the Product is 25168, the Content of the fmall Trapezia, *b*.

Subtract 826 from 1120, the Remainder is 294, the Length of *c*; and becaufe both the Offsets are alike, multiply 294 by 36, the Length of the perpendicular Offset, the Product is 10584, the Content of the fmall Piece, *c*.

In

In the same manner deal with the rest of these small Pieces round the Field, and set down the Product of each in an orderly manner one under another; so shall the Sum give the exact Content of these small Pieces, which added to that within the Station Lines, gives the true Content of the Field in square Links, which reduce into Acres, &c. as before directed.

Note, The Performance of this being tedious, I shall in the next Chapter lay down a more Practical Method for casting up the Content of any Piece of Land.

CHAP. III.

Shewing the Use of the Theodolite in Surveying Land by the help of the Needle and Limb together.

SECT. I.

N this Method of Surveying Land, the Angle which every Line makes with the Meridian is measured by the Limb of the Theodolite, and therefore much preferable to that before mentioned in the aforegoing Chapter by the Needle only, because the Degrees and Minutes are better estimated on the Limb of the Instrument that 'tis possible they should be in the Box of the Circumferentor; and this gives the Theodolite the preference to any other Instrument, because we can work by the Limb only, without regarding the Needle at all; but if it be more convenient to make use of the Needle, we may do it in the following manner, being the most exact and absolute Method yet known for Surveying large and spacious Tracts of Land.

For the Needle being observed to play well, when it hath settled in the direction of the Meridian and is at rest, the Box may be moved round the fixed Center by turning the Index on the Limb and the Point mark'd with 360 in the Box, brought directly against the north End of the
Needle,

Needle, with greater Exactnefs than a degree, and its Parts can be eftimated in any other Part of the Box; befides we have this Advantage which is very confiderable, that we can make ufe of a fhort light Needle whofe friction being lefs, plays better than a longer and heavier.

Let the Lines *o a, b c, d e, f,* in *Fig.* 17. reprefent the Station Lines near the Bounders of a Field; then the Angle which each makes with the Meridian may be obferved in the following manner.

Firft, having fet up a Mark at *o,* Meafure forwards with the Chain on the Line *o a* to *a* 600 Links.

Plant the Inftrument at *a,* and bring the Index to 360 on the Limb, and turn the whole Inftrument about (whilft the Needle hangs in the direction of the Meridian) till 360 in the Box is brought directly againft the north End of the Needle, and there fix the Inftrument, then is the Telefcope fet in the direction of the Meridian alfo; and in this Pofition is the Inftrument to be planted at every Station.

Now turn about the Index till the Hair in the Telefcope cuts the Mark left at *o,* and note in the Field Book the Degrees and Minutes which the Index cuts on the Limb, *viz.* 207 Degrees 20 Minutes, being the Quantity of the Angle which the Line *a o* make with the Meridian.

Remove the Inftrument from *a,* leaving a Mark at that Station and proceed with the Chain to *b,* and there plant the Inftrument, then bring the Box exactly to the north End of the Needle as aforefaid, and direct the Telefcope to the Mark left at *a,* and note the Degrees and Minutes cut on the Limb by the Index, *viz.* 285 Degrees 10 Minutes, which is the bearing of *b a*

or

or Quantity of the Angle which that Line makes
with the Meridian.

It would be Tautology to repeat the manner
of meafuring the other Lines and Angles in this
Figure, but obferve that when the Inftrument is
fixed in the direction of the Meridian, we fre-
quently obferve the Needle by moving it from
the Point at 360 with a Knife, &c. then if it fwings
backward and forward freely without jogging or
ftopping, and fettles again to 360 exactly; we
may conclude the Inftrument is right in the
direction of the Meridian to make an
Obfervation.

It you fufpect the Needle to be acted upon by
fome hidden magnetick Power, as when you are
Surveying in mountainous Lands, where there
may poffibly be Iron Mines in the Earth, which
will attract the Needle, you may obferve whether or no it be drawn afide in the following
Manner.

As when the Inftrument was planted at *e*, the
North End of the Needle pointing to 360 in the
Box; after the bearing of *e d* was noted, direct
the Tellefcope forwards to *f*, and note the Angle
which the Index cuts on the Limb; *viz.* 200° 5′o,
then the Inftrument being planted at *f*, becaufe
the bearing of *e f*, obferved at *e*, is more than
180°; fubtract 180 there-from and to the
Remainder 2°o 5′o on the Limb, fet the Index
exactly; but if the bearing of *e f* had been lefs
than 180, add 180 thereto, and to that Number,
being the Index on the Limb, now turn about
the whole Inftrument till the Hair cuts the Staff
left at *e*, and then, if the *North* End of the Needle
points to 360, as at the laft Station, the Bearing
of that Line is truly obferved.

For the magnetick Power that attracts the
Needle,

Needle, being fuppofed at a great Diftance, the Direction on fuch a Piece of Land as is commonly furveyed by the Theodolite, will be the fame: But if the attractive Power be near the Inftrument, the Needle will incline thereto.

Now follows the manner of protracting thefe Obfervations,

Lines	Links	Sta.	Deg.	Min.
a, o,	600 ——— a,		207	20
b, a,	500 ——— b,		285	10
c, b,	1000 ——— c,		190	00
d, c,	500 ——— d,		91	55
e, d,	500 ——— e,		125	20
f, e,	1600 ———f,		20	50
o, f,	500 ——— o,		289	15

SECT. II.

A new Method of protracting any Obfervations made in the Field by the Needle.

BY which a Plan may be drawn on the Paper from one Meridian only, and all the Angles therein laid down by once applying the Pro-tractor to that Meridian by the help of a parallel Ruler, being very exact and expeditious.

Provide a circular Protractor, whofe Numbers encreafe the fame way as on the Limb of the Theodolite, and a parallel Ruler of a convenient Length, then draw a Right-line N S, *Fig.* 17. (with a black lead Pencil) for a Meridian, and affign a Point therein, as at *o*, to which Point apply the Center of the Protractor, and turn it about till the Diameter lyes on the Lines N S, with 180 towards N, (that part of the Limb of the

Theodolite

Theodolite being always kept Northward of the Field.)

The Protractor held in this Pofition, lay the Field Book before you, and againſt 207 Degrees 20 Minutes, the bearing of the Firſt Line *a o* cloſe to the Limb of the Protractor, make a Mark with the protracting Pin or Needle, and cloſe to that Mark write *a* with a black lead Pencil.

Hold the Protractor in the fame Pofition, and againſt 285 Degrees 10 Minutes, the bearing of the next Line *b a*, make a Mark with the protracting Pin, and cloſe to that Mark ſet *b*.

In the fame manner keeping the Diameter of the Protractor cloſe to the Meridian as it was at firſt laid; make a Prick with the protracting Pin, cloſe to the Limb of the Protractor, againſt the bearing of each reſpective Line as noted in the Field Book, and cloſe to each Prick ſet the Letter or Number of that Line; ſo againſt 190 Degrees the bearing at *c*, make a Prick and write *c*, againſt 91 Degrees 55 Minutes write *d*, againſt 125 Degrees 20 Minutes write *e*, *&c. Vide Fig.* 17.

Having mark'd the bearing of each Line round the Protractor, lay it afide, and apply the edge of your plotting Scale to *o* at the Center, and *a* mark'd by the Limb of the Protractor; the beginning of the Numbers coinciding with *o*, and encreaſing towards *a*, and prick off 6 Chains the Length of the Line *o a*, and with Ink draw the Line *o a a*.

Lay the parallel Ruler to the prick'd Line *p o b r*, ſo that the edge cuts the central Point at *o*, and the Point at *b*, as mark'd by the Limb of the Protractor; and move it parallel till the Edge cuts the Point at *a* in the Line *p a b r*, and with the Point of your Compaſſes draw the occult Line *p a b r* by the Edge of the parallel Ruler;
                                                            then

then becaufe the Length of the Line *a b* is 5 Chains, lay the plotting Scale to *a* and Prick off 5 Chains, and draw the Line *a b*.

When you had drawn the occult Line *p a b r*, through the Point *a*, you might fet *a b* thereon towards *p* as well as towards *r*; but if you obferve in what Direction the Letter *b*, as mark'd by the Limb of the Protractor, ftands from the central Point *o*, in the fame direction muft the Line *a b* be fet from the Station Point *a*; alfo when the Ruler is laid to the Station *b*, you cannot be at a lofs whether you fhould draw the Line *b c* upwards or downwards, if you obferve in what direction the Letter *c* ftands from the central Point *o*; therefore in the fame Direction draw *b c* from *b*, or the Angles mark'd external in the Field Book will be a fufficient Direction.

Lay the parallel Ruler to the central Point *o*, and the Mark at *c*, and move it parallel in that Direction, till the Edge cuts the Point *b* at the end of the Line *a b*, and by the Edge of the parallel Ruler, draw an occult Line, fetting thereon from *b* 10 Chains, and draw the Line *b c*.

Again lay the Edge of the parallel Ruler to the Point at the Center of *o*, and to the Mark at *d*, and move it up to *c*, and draw *c d*.

In the fame manner deal with the other Lines and Angles, fo fhall the laft Line *f o* cut through the Point *o*, and its Length be 5 Chains, as noted in the Field Book, which proves the Plott to be truly laid down.

In thefe Obfervations the Station Lines only are inferted, the Offsets from thence to the Bounders are omitted, the manner of plotting them being already laid down before.

When the Bounders of the Field are drawn, and the Name thereof entred in the middle of the

Plott,

Plott, you may with a piece of Bread rub off the Marks that were made with the Pencil round the Edge of the Protractor and meridian Line, fo will the Plott be ready for cafting up.

But if feveral Fields are to be plotted to-gether, you muft draw a Line through the firft Station Point in each, parallel to the Meridian in the firft Plott, from which the Plott of each Field may be laid down in the fame manner as *Fig.* 17.

Obferve, neither the Circle nor Figures, ex-preffing the Angle which each Line makes with the Meridian, are ufed in Practice, though in-ferted in the Scheme to demonftrate the Nature of the Work; alfo if you lay the Edge of the thin plotting Scale clofe to the Edge of the parallel Ruler, and move it forwards on the Paper with the parallel Ruler, till the Edge of the Scale cuts the Point at *a*, and bring the beginning of the Numbers on the Scale to the Point *a*, you may draw the Line *a b* by the Edge of the Scale held in that Pofition to 500, the Length of the Line without drawing any other but the Station Line it felf.

SECT. III.

A New Method of calculating or cafting up the Area of a Plott of Land in Acres, &c.

ACcording to the Rules before mentioned in Chap. 1. the whole Plott muft be reduced into Trapezias and Triangles, and the Length of each Bafe and Perpendicular meafured by the Scale; but fince it is often neceffary to lay down the Plott by a fmall one, as $\frac{1}{4}$ of an Inch or lefs; if you err 8 or 10 Links in taking off the Length of the Bafes and Perpendiculars (which may eafily happen if the Lines be not drawn very neat and
fmall)

fmall) and there being feveral fuch Bafes and
Perpendiculars, the Error may be confiderable in
the whole Plott, and then alfo the Bafe and
Perpendicular of each of thefe Triangles muft be
multiplied together feverally, and their Products
added together for the whole Content.

Whereas by this Method the whole Plott, (let
it confift of many Sides or few) is caft up by
applying the Scale but to one Bafe and one
Perpendicular, and confequently by one Multiplication, and the Truth of the Work is demonftrated by the firft Theorem in Chap. 1. *viz.* That
Parallelograms (and confequently Triangles)
conftituted upon the fame Bafe, and between the
fame Parallels are equal.

Let the four-fided Figure *a b c d*, *Fig.* 18. be
reduced to a Triangle, whofe Area fhall be equal
to that of the four-fided Figure.

Firft extend one of the Sides as *c d*, then lay
the parallel Ruler to the Points *a* and *d*, and
move it parallel till the Edge cuts the Point *b*,
then by the fame Edge make a Mark in the
extended Line *c d* at *e*: Laftly lay a ftrait Ruler to
the Points *e* and *a*, and draw the Line *e a*, fo fhall
the Area of the Triangle *a c e*, be equal to the
Area of the four-fided Figure *a b c d*.

For the Triangles *d o e*, and *b o a*, having
Bafes of the fame Length, any lying between
the fame Parallels are evidently equal; then if
the Triangle *b o a* is left out of the four-fided
Figure *a b c d*, and the Triangle *d o e*, taken in,
and the Areas of thefe two Triangles being
equal; it follows, that fuch an equal Quantity
of Space is left out in one part of the Figure as
is taken in on the other, and the Area muft ftill
be the fame.

Again, let *Fig.* 19. be reduced into a Triangle.

angle.

Firſt extend the Line *f o*, and apply the parallel Ruler to the Points *o* and *b*, and move it up parallel to the Point *a*, and where the Edge cuts the extended Line *f o*, make a Mark at *g*, then lay the Ruler to the Point *g* and *c*, and move it up to *b*, and make a Mark in the extended Line or Baſe at *h*.

Lay the Ruler to the Points *h* and *d*, and move it to *c*, then make a Prick in the Baſe at *i*.

Lay the Ruler to the Points *i* and *e*, and move it to *d*, and make a Mark in the Baſe at *k*.

Laſtly draw the Line *k e*, ſo ſhall this ſeven-ſided Figure be reduced to a three-ſided one whoſe Areas are equal, ſo may the Triangle *f e k* be caſt up by one Multiplication only.

But Note, inſtead of laying the Ruler to the Points *i* and *e*, if you had laid it to *d f*, and moved it up to *e*, and drawn the Line *z d*, the Triangle *z d i*, would have contained the ſame Area as *f e k*, and this often is neceſſary to prevent the Sides of the reduced Triangle being extended too long, and making the Angles thereof too acute.

Apply the ſame Scale by which the Plott was laid down to the Baſe, and meaſure its Length, alſo meaſure the Length of the Perpendicular; multiply theſe two Sums together; the half of their Product is the Content of the Plott in ſquare Links, which reduce into Acres, &c. as before directed.

Alſo obſerve that we commonly chuſe to extend one of the ſhorteſt Sides of the Plott to be the Baſe of the Triangle, as the Side *f o* which we draw with a black lead Pencil as *o k*, as well as *k e*, *i d*, or *z d*, and rub em off again with a Piece of Bread, as ſoon as the Content of
the

the Field is entred with its Name in the middle thereof.

If in ufing the parallel Ruler at the firft Tryals you find it apt to flip on the Paper, which you may do if you be not very careful to hold it clofe down thereto, that Inconvenience may be prevented, if you make ufe of three fmall Pins or Needles, thus: Stick the three Pins in the three Firft Angles, as at *o*, *a*, and *b*, then apply one of the inner Edges of the parallel Ruler, to the firft and third *o* and *b*, and move the other inner Edge to the Second at *a*, take out the Pin at the Second, and put it in the Bafe or Line extended where the Ruler cuts it at *g*; again lay the Ruler to this Pin at *g*, and to another at the Fourth Angle at *c*, and move the Ruler to the Fifth Angle at *d*, take out the Pin at *d*, and ftick it in the Bafe at *h*, and proceed in this manner with the reft till the Plott is reduced.

SECT. IV.

Shewing how to reduce the irregular Bounders of a Field to ftrait Lines, on order to find the Area thereof.

LET *a b c d e f g h i k*, *Fig.* 20. reprefent the Bounders of a Field, whofe Content is defired.

Firft, produce fome one of the longeft Sides as *i k*, then lay the parallel Ruler from the Angle *i* to *g*, the next but one, and move it up to the Point *h*, and where it cuts the Line produced, make a Mark at *r*, and draw the ftrait Line *r g*, and it will reduce that Side of the Figure bounded by the two Lines *i h*, and *h g*, to another bounded by *r g* one Line only.

In like manner *r g* being produced, and the
parallel

parallel Ruler laid from *g* to *e*, and moved up to
the Angle *f*, the Edge cuts the extended Line *r g*,
at *y*; Secondly, lay the Ruler from *y* to *d*, and
move it up to *e*, it cuts the extended Line *e g* at
z; Thirdly, lay the Ruler from *z* to *c*, and move
it up to *d*, and where it cuts the extended Line
r g, make a Mark at *x*; Laſtly, draw the ſtrait
Line *y c*, ſo ſhall the Side *g c* which conſiſted of
the four Lines *g f*, *f e*, *e d*, and *d c*, be reduce to
the Side *y c* conſiſting of one Line only, and in
like manner might we proceed, if the Lines were
never ſo many, ſo may the ten-ſided Figure be
reduced to a four-ſided one, and then to a
Triangle which may be caſt up by one Multiplication only.

This is the ſame Method before laid down for
reducing a many ſided Figure to a Triangle, but
if you have not a parallel Rule, do thus:

Having produc'd the Side *k i*, lay the Edge of
a ſtrait Ruler from *i* to *g*, then take with a pair of
Compaſſes the Diſtance from *h* to the Edge of the
Ruler, and with this Diſtance let one Point of the
Compaſſes move gently cloſe to the Ruler, while
the other traces out a Line parallel to it, and
croſſes *k i* at *r*, and draw *r g* as before.

In the ſame manner deal with the other Sides,
uſing the Compaſſes in this manner inſtead of a
parallel Ruler.

Provide a plate of thin Braſs in form of an
Arch of a Circle, near whoſe ends let there be
drill'd ſmall Holes, through which ſtring it with a
very fine Hair; and then an Hedge as *g c*, *Fig.* 20.
bends in and out in ſeveral Places, and thoſe
Bends contain very ſmall Spaces; lay the Hair
over it length-ways ſo that the Quantities cut off
from the Figure thereby, may be equal to thoſe
added to it, and with a protracting Pin near the
ends

ends of the Hair, make two Marks, through which, draw a ftrait Line, and fo will this irregular Side be reduced to a regular one; and here it may be obferv'd that in very fmall Bends by the Eye, you may judge better than by the Compaffes.

But if Hedges confift of large Curvatures, chufe out fuch Points, and fo many of them that Right-lines drawn from Point to Point may vary the Quantity by fuch Quantities only as may be rejected, and herein the Hair will be a Ready Affiftance.

SECT. V.

The manner of reducing hypothenufal to horizontal Lines.

WHEN we meet with an Hill in Surveying a Piece of Land; we can only meafure the hypothenufal or flope Lines thereof, on the Superficies of the Hill, which being confiderably longer than the Bafe or level Lines on which the Hill is fituated, as Lines *a b*, *b c*, *Fig.* 21. are longer than *a o*, *o c*, therefore when we plott this Hill (becaufe we cannot make a convex Superficies upon a Piece of plain Paper) we muft reduce the hypothenufal to horizontal, Lines that all the Lines in the Plott may be laid down alike in *Plano*.

For the Lines of level only muft be exprefs'd in a Plott; that every Field therein may lye in its true Situation; for if *a b*, and *b c*, were laid down on Paper as meafured in the Field, they would reach to *d*, and not only thruft the next Hedge out of its true Pofition, but alfo take up a great Space in the next Field, making that too little.

Let

Let *Fig.* 21. reprefent an Hill; at the foot of which the Theodolite is planted, which being fet level in order to meafure the Angle at *a*, the Telefcope when directed towards *b*, at the top of the Hill, cuts the Ground; therefore take the Pin out of the Quadrant, and elevate the Telefcope to the Mark at *b*, (which muft be fet the fame Diftance from the Ground as the Telefcope is) and when the Hair cuts the Mark at *b*, the Index fhews the horizontal Angle on the Limb, and the Quadrant the Angle of Elevation *b a o*, 25 Degrees 50 Minutes both at the fame time, which note in the Field Book one over againft the other.

The Inftrument removed from *a*, and planted level on the top of the Hill at *b*, the Telefcope when directed towards *c*, cuts the Element, therefore take out the Pin from the Quadrant, and deprefs the Tellefcope to the Mark at *c*, and then the Quadrant will cut 21 Degrees 34 Minutes, and the Length of *a b*, as meafured up the Hill by the Chain is 1200 Links, and *b c* 1416.

In order therefore to plott thefe Obfervations, firft, draw the Right-line *a d*, but do not fet the Length 1200 Links thereon, becaufe the Angle of Elevation is noted in the Field Book againft the horizontal Angle, which fhews that this Line is to be reduced to a Level; therefore lay the Center of the Protractor to *a*, the Diameter coincident with *a d*; and againft 25 Degrees 50 Minutes, the Angle of Elevation, make a Mark, and through it draw the obfcure Line *a b*, fetting thereon 1200 Links the Length of the Hypothenufe, at the End of which make a Mark at *b*.

Having drawn the Angle of Elevation *b a o*, take a fquare Protractor or any other Square that
hath

hath one Right Angle, and two ſtrait Edges and apply one Edge thereof to the Right Line *a d*, whilſt the other Edge cuts the Point *b* in the obſcure Line *a b*, and thereby let fall a Perpendicular from the Point *b*, which falls on the Line *a d* at *o*, ſo ſhall the Line *a o*, be the true horizontal Line which muſt be laid down in the Plott.

In the ſame Manner, reduce the Hypothenuſe *b c*, by firſt drawing the Angle of Depreſſion *d o e*, 21° 3′ 4 ſetting the Length of the Hypothenuſe *b c* 14, chain 16 Links on the obſcure Line *o e*, and where the Length 1416 Links reaches from *o*, make a Mark at *e*. Laſtly, from *e* let fall a Perpendicular on the Line *o d*, which falls at *c*, ſo ſhall the Line *o c*, be the true Horizontal Line.

Or elſe having noted the Quantity of the Angle of Elevation, and Length of the Hypothenuſe in the Field Book, you may find the horizontal Line by the help of the following table.

A Table ſhewing how many Links to deduct out of every Chain's Length in the Hypothenuſal-Line.

Deg.	Min.	Lin.	Deg.	Min.	Lin.	Deg.	Min.	Lin.
4	3	$\frac{1}{4}$	19	57	6	29	32	13
5	44	$\frac{1}{2}$	21	34	7	30	41	14
7	1	$\frac{3}{4}$	23	4	8	31	47	15
8	6	1	24	30	9	32	52	16
11	29	2	25	50	10	33	54	17
14	4	3	25	50	10	34	55	18
16	16	4	27	8	11	35	54	19
18	12	5	28	21	12	36	52	20

Having

Having the Angle of Elevation 25 Degrees 50 Minutes, and the Length of the Hypothenufe *a b*, 12 Chains given thence to find the Length of the horizontal Line.

Look in the Table for 25 Degrees 50 Minutes, and againſt it you will find 10 Links, and ſo many muſt be deducted out of every Chain in the Length of the Hypothenufe, then if 1 Chain or 100 Links requires 10 Links to be deducted from thence, 12 Chains or 1200 Links, requires 120 Links to be deducted; therefore ſubtract 120 Links from 1200, the Remainder is 1080, the Length of the horizontal Line *a o*.

Again, the Angle of Depreſſion at *b*, is 21 Degrees 34 Minutes, and the Length of the Hypothenufe or ſlope Line *b c* 1416 Links, you will find in the Table againſt 21 Degrees 34 Minutes 7 Links, then if 100 : 7 : 1416 : 99, therefore ſubtract 99 Links out of 1416 the Length of the ſlope Line, the Remainder is 1317 Links, the Length of the level Line *o c*; But if you cannot find the given Angle of Elevation in the Table, make uſe of that which approaches the neareſt thereto; and *Note*, Surveyors in Practice ſeldom take notice of a gradual Aſcent, if it does not make an Angle of above 5 or 6 Degrees or thereabouts, the difference between the ſlope and level Line, being then inconſiderable, except in ſome extraordinary Caſe, and then 'tis ſafeſt to make uſe of the firſt Method here laid down, becauſe the Table is too ſhort, but if you have a correct Table of Sines and Logarithms, you may make uſe thereof.

If you are working with the Chain, and would find the horizontal Line of an Hill, you may carry a ſmall Quadrant in your Pocket, with which meaſure the Angles of Altitude, and note

it

it in the Field-Book againſt the Chord or Sextant of the horizontal Angle obſerved at that Station, (but let the Mark be ſet the ſame Diſtance from the Ground with your Eye when you obſerve the Angle of Altitude) and proceed to reduce the Line as aforeſaid.

Alſo you may obſerve an Angle of Altitude if you have only the plain Table in the Field, by turning it down into the notch of the Ball and Socket, making it ſtand Perpendicular by applying the String and Plummet thereto, and then the Index and Sights ſcrewed to the Center of the Table may indifferently ſerve the turn, but a Quadrant is better.

You muſt ſhade over that part of your Plott where the Lines are thus reduced with the Re-preſentation of Hills, leſt another Perſon ſhould meaſure them by the ſame Scale with the other Lines, and find them to differ.

If a Field have the bottom and top Lines level, and Sides riſing alike, it is to be accounted but as a declining level, and to be meaſured as a Level Ground in regard of the Quantity of Superficies, though the ſide Lines muſt be reduced to make a regular Plott in reſpect of the adjacent Fields that are level; but if a Ground be level at one End and both Sides, and an Hill riſing up along the Middle, or if there be ſeveral Hills in the Middle, thereof the Superficies will be more than in a Plain bounded by the ſame Limits.

Now Surveyors differ in their Opinion, in re-ſpect of caſting up the Content of ſuch a Field; ſome argue on the Tenant's Behalf, then ſince all Vegetables ſtand in a Perpendicular on the Earth; (that is, grow ſtrait upwards) as much will grow on the horizontal Line as on the Hypothenuſe,

and

and therefore the Lines ought to all be reduced to a level, and the Content to be deduced from the Plott fo laid down.

Others fay that there ought to be marks placed on the top of the moft remarkable Hills, and the Chain drawn over Hill and Dale, and the flope Lines laid on the Paper of the fame Length as meafured in the Field, and the true Content in Acres, &c. deduced from thence although the Slopes be reduced afterwards, that the Field may be laid in its true Situation in refpect of others adjacent in the fair Plott.

'Tis hard to determine which way is to be practiced in all Cafes; for though by the laft Method you will have the true Quantity of Superficies more nearly given, yet the allowance in the firft is often but reafonable, if the Soil of the Hills is not fo profitable as if the whole Field was fituated on a Plain, but the Reader may ufe which he fhall think moft proper.

CHAP. IV.

Shewing how to Survey and make a perfect Draught of several pieces of Land lying together as a Manor, &c. Also how to compare the Bearing and Angles one with another, at each Station, as observ'd by the Theodolite, in order to correct any Error that may arise in measuring the Angles in the Field as well as protracting them on Paper.

SECT. I.

HAVING in the former Chapters laid down the best and most practical Methods for measuring any Piece of Land by the most proper Instruments, I here subjoin the manner of Surveying several Parcels lying together; an Example of which may be taken from the small Tenement or Farm, *Fig.* 22.

First I take a View of the Land, considering at which

which Part thereof it will be moſt convenient to
begin, and proceed with the Work; and becauſe
'tis beſt working in a Lane as often as Oppor-
tunity preſents; therefore I ſet up the Theodolite at ⊙ 1 in *Charlton* Field.

Then I enter in the Field-Book the Title of
the Survey, and in the middle Column ⊙ 1, and
then ſend a Station Staff forwards in the Lane, as
far as I can ſee diſtinctly, (the farther the better)
as to ⊙ 2, (and when the Station Lines are within
the Fields, I ſend the Staff to the next eminent
Bend in the Hedge, or even to the farther end
thereof, if the Line from the Inſtrument to the
Staff be not at too great a Diſtance from the
Hedge, ſo as to cauſe Offsets greater than a
Chain or a Chain and half, or thereabouts, for
Offsets taken too long are not ſo eaſily laid off at
right Angles from the Station Line) and to that
Staff at ⊙ 2, I direct the Teleſcope, and note the
Degrees in the Box cut by the north End of the
Needle, *viz.* 356 Degrees 10 Minutes, which I
enter in the Field-Book for the Bearing of this
firſt Station Line.

Then I ſet up the Staff in the Hole over which
the Center of the Inſtrument was plac'd, to
which Staff I direct one of my Aſſiſtants to apply
the Ring at one End of the Chain, whilſt the
other Aſſiſtant ſtretches it out in a right Line
towards ⊙ 2, letting it lye on the Ground in that
Direction; till the Occurrences in this Chain's
Length are entred in the Field-Book, *viz.* I
meaſure the Diſtances of the Chain from the
Bounders of each Field, which I enter in the
Columns or Offsets, that on the right Hand of
the Chain in the right Hand Column, and that on
the left in the left Hand Column; and if the Land
is Part of that which I am about to ſurvey, I
write

write in one of the outfide Columns by what
Name it is call'd, but if it belongs to a Stranger, I
write the name of the Owner thereof, and in all
Cafes exprefs to which Land the Hedge belongs.

So at the Length of 20 Links from ☉ 1, I lay
the Offset Staff at right Angles with the Chain,
and meafure the Diftance from thence to the
corner of Turfy Leas, which I find to be 15
Links; therefore in the middle Column reprefenting the Station Line I write 20, and againft it
in the right Hand Column of Offsets I enter 15;
likewife when I come to 40 Links in the Chain-
Line I am againft the Corner of the Cow-pafture;
therefore, I lay the Offset Staff to the Chain, and
meafure the Diftance from thence to the Corner
of the Cow-pafture 80 Links, which I enter in the
left Hand Column of Offsets againft 40 in the
middle Column, denoting that at the Length of
40 Links from ☉ 1 the Offset, 80 Links reached
the Corner of Cow-pafture on the left Side of the
Station-Line.

The Hedges on each fide the Line, running
on very nearly ftrait from thefe Corners, I take
no more Offsets in this firft Chain's Length nor at
the fecond; but when I have laid the Chain a
third time, and come againft 80 Links, I there
take an Offset on the left Hand of the Chain
Line, becaufe the Hedge varies its Direction,
making a confiderable Bend; for though the
Diftance from the Chain to the Hedge continually varies from the Corner to this Place; yet I
only take Offsets at each End, omitting the
intermediate Parts; fince when the extreams of a
right Line are given, that right Line is alfo
given, but when the Hedge runs on with a continued but irregular Curviture, then I take Offsets
at every Chain or half Chain's Length, or oftner

as

as the Thing requires.

In this manner I proceed with the Chain till I come to the Staff at ⊙ 2, obferving as I go along the Bends in the Fences on each fide of the Lane, to every one of which I take an Offset, writing the Length of each on the right or left fide of the middle Column in the Field-Book reprefenting the Station Line, according as they were laid off in the Field.

Note, the Mark ⊙ in the Field-Book denotes a Station; *B* a Bearing, < an Angle, *cu.* the cutting of an Hedge by the Chain, *a g*, fome remarkable Object on the farther fide of the Hedge, as another Fence fhooting up thereto; *ret.* return to a former Station, *&c.*

Being come to ⊙ 2, I there plant the Inftrument, and fend the Station Staff forwards in the Lane as far as I can fee it, as to ⊙ 3, and then bring the Index to 360 on the Limb, and turning the whole Inftrument about I direct the Telefcope to the Staff left at ⊙ 1, and there fix the Inftrument; and then turn about the Index on the Limb, till through the Telefcope I fee the Staff at ⊙ 3, and then find that the north End of the Needle points at 338 Degrees, and the Index cuts on the Limb 161 Degrees 50 Minutes; therefore under ⊙ 2, in the middle Column of the Field-Book, I enter 338 Degrees, and under that 161 Degrees 50 Minutes, denoting that at the fecond Station the Bearing of the fecond Length is 338 Degrees, and the Angle which the Index cuts on the Limb is 161 Degrees 50 Minutes.

The Rule I obferve in meafuring each Angle is this; firft I bring the Index to 360, and with that part of the Limb towards me, I direct the Telefcope to a Mark at the laft Station, and there fix the Inftrument; then I turn about the Index

on

on the Limb, till I fee the Hair in the Tellefcope cut a Mark at the next Station before me, fo fhall the Needle fhew the Bearing of the next Line, and the Index on the Limb fhews the Quantity of the Angle at the prefent Station.

The Angles and Bearing of the Lines are taken at once fetting the Index, as eafily and ex-peditioufly as the Angle it felf only; therefore infert the Bearing of each Line in the Field-Book as you fee in the Form thereof; for then you may prove the Truth of your Work in the Field at each Station, before you leave it, by one of the following Rules.

If to the prefent Bearing, be added 180 Degrees, and from the Sum you fubtract the laft Bearing, then the Remainder will be the prefent Angle.

Of if to the prefent Angle, you add the laft Bearing, and from the Sum fubtract 180, then will the Remainder be the prefent Bearing.

But if the Degrees to be fubtracted are more than thofe from which they are to be fubtracted, the latter muft be encreafed by 360, and then fubtract. And if the Remainder be more than 360, then abate 360, and the Refult gives the Degrees required.

So at ☉ 2, if to the prefent Bearing 338 00′, you add 180°, the Sum is 518°, 00′ from which Sum, if you fubtract the laft Bearing at ☉ 1, 356° 10′, the Remainder is 161° 50 equal to the prefent Angle.

Likewife, if to the Bearing at ☉ 3 1° 30′, you add 180 Degrees, the Sum is 181° 30′, which is lefs than 338 00′, the Bearing of the laft Station, therefore 181° 30′ muft be encreafed by 360, and then the Sum is 541° 30′, from which if you fubtract 338, the Bearing of the laft Station, the Remainder will be 203° 30′, equal to the prefent Angle.

In like Manner may any other Angles be exa-mined,

mined, and if found erroneous, the Error may be correcſted, before 'tis communicated to the following Part of the Work.

Therefore when you have noted the Bearing of the preſent Station, write it in one of the outſide Columns of the Field-Book, and adding 180 thereto, ſubtraſct the Bearing at the laſt Station there-from; and then, if the Angle thus calculated from the Bearings, doth agree with that which the Index cuts on the Limb, you may conclude the Angle is rightly obſerved, and therefore may be entred in the Field-Book.

But obſerve, tho' the Numbers thus compared will be very nearly alike, yet ſometimes they may differ ſome few Minutes, becauſe the Diviſions in the Box being ſo much ſmaller than thoſe on the Limb, the Degrees and Minutes can't be eſtimated alike in both; but yet you will be ſure always to correſct and avoid any groſs Error before you proceed with the following Work; and to this End the before-mentioned Rules are of excellent Uſe.

Theſe Direſctions I ſhall not repeat, tho' I make Uſe of the them throughout the whole Work, unleſs any thing new occurs in meaſuring of the other Lines and Angles, referring the Reader rather to the Field-Book and Plan of the Work, than tiring him with Repetitions.

From ⊙ 2: I proceed with the Chain towards ⊙ 3: but at twenty Links in the firſt Length from ⊙ 2, I am againſt the Hedge that parts Homecloſe from Turfy-Leas; therefore I take an Offſet thereto perpendicular from the Chain Line, and enter in the Field-Book *a g* 17 Links, and this will hereafter be of Uſe in cloſing the Plott.

Being come to ⊙ 3, I there obſerve and prove the bearing and Angle at that Station, and then
proceed

proceed with the Chain towards ʘ 4; but firſt at ʘ
or the preſent Station, I meaſure an Offset to the
right 10 Links, and to the left 20 Links; at 41 in
the Chain Line; I am againſt the Orchard Hedge
at 204, the Orchard Pales at 261, I am againſt the
Gate that leads into the Yard, and alſo againſt
another that goes into Cow-Paſture; therefore to
each of theſe Remarks I meaſure an Offset from
the Chain Line, and enter them in the Field-
Book.

In going from ʘ 4 to ʘ 5, the Chain touches
the Brow of the Ditch at 2 Chain 20 Links from
the laſt Station; therefore againſt 220 in the
Field-Book I write *0*, denoting that there was *0*
or no Diſtance from the Chain to the Ditch, and
by the Brow of the Ditch is meant the determined Diſtance of 5 Links from the Stem of the
Hedge.

Being come to ʘ 6, I ſend a Staff to the farther
Side of the Field called the Stockin, and if I cannot ſee the Mark through the Hedge, I cauſe the
Bows to be removed or held back till I can plainly ſee the Mark, and if that can't be done I make
my Station a little on this Side, or elſe beyond
ſuch a thick Place till I can plainly ſee the Mark,
and draw the Chain through the Hedge in a ſtrait
Line, and where it cuts the Hedge I write *cut*, as
here it does at 10 Links; but if the Fence is a
Wall I allow for the Thickneſs thereof, and
always meaſure the neareſt Diſtance between
Station and Station that can poſſibly be.

Being come to ʘ 9 in the Stockin, I cauſe a
Staff to be ſet up in the very Corner of the Field
next the Lane where the Bounders meet, to
which I meaſure from ʘ 9; ſo ſhall the End of this
Line coincide with the Offset which I took to
this corner from the Station Line in Lane, which
                                                              will

will be a Proof that the Work is truly laid down when I come to protract it.

If upon a Piece of Paper with the Pen only you make an Eye-Draught of the Lane, and that Field which you have finifhed, fetting your Station Lines with their Numbers, as you made them in the Field, you will plainly fee your Work as you go along, and be able to diftinguifh which Bounders of the prefent Field are already obferved in the Precedent, as well as be directed, with a great deal of Eafe, how to proceed with your Work, when you come to protract it.

Having made all neceffary Obfervations round the *Stockin*, I return to ☉ 9, and with 360 on the Limb towards me I direct the Tellefcope to a Mark, at the laft Station before I came to this, *viz.* ☉ 8, and fixing the Inftrument there, I next direct the Tellefcope to 10, and note the Angle and Bearing at that Station; then I lay the Chain through the Hedge from ☉ 9 towards ☉ 10, and take an Offset to the Fence where each Partition Line joins it on the other Side, by the Help of which, together with the other Offsets on the fur- ther Side in the Lane, moft of the inward Fences of the Orchard, Garden, *&c.* may be drawn.

From ☉ 10 I cannot fee into the very Corner next the Lane, therefore I take an Offset thereto 20 Links from the Station Line, and fo are the out Lines of the Garden, Yard, *&c.* finifhed.

Then I return to ☉ 8 in the *Stockin*, and here it may be obferved, that when I defign to return to any Station, before I leave it I cut up the Turf with a little Paddle, which I fix in one End of the Offset-Staff, or make fome fuch Remark that I may be fure readily to find the Place in which the Station Staff before ftood; and in the Field-Book to this Mark ☉ I write return; then planting the

Inftrument

Inſtrument at ⊙ 8, I direct the Tellescope to the Mark left at ⊙ 7, and here alſo I obſerve as a general Law, to obſerve the Angle with that Line which was meaſured immediately before I came to the Station where I took the Angle the firſt time: So here I obſerve the Angle made with ⊙ 7, ⊙ 8, and not with any other, as ⊙ 8, ⊙ 9; therefore according to this conſtant Law I direct the Tellescope back to ⊙ 7 in the *Stockin*, and fixing the Inſtrument there, I next direct the Tellescope to ⊙ 11, in *Home-Cloſe*, and *Note* the Angle, *&c.* as in the Field-Book.

After I have meaſured the Angle *&c.* at ⊙ 11, and am going forwards towards ⊙ 12, at 76 Links of the Chain I perceive my ſelf over againſt the Fence that parts *Out-Wood* from *Crab-tree-Cloſe*, I therefore ask the Follower of the Chain, how many Arrows he hath in his Hand, he anſwers 4, therefore I enter in the middle Column of the Field-Book 476, and againſt that on the left I write 61, the Length of the Offset denoting, that at the Length of 4 Chains 76 Links from ⊙ 11, I laid off an Offset to the left, 61 Links, and proceed to obſerve and enter in the Field-Book the reſt of the Occurrences round *Home-Cloſe*, cloſing it at the End of the Line from ⊙ 12 next the Lane.

Then I return to ⊙ 12, and proceed to ⊙ 13, cloſing *Turfey-Leas* at the Corner; to which I took the firſt Offset from ⊙ 1, in Charlton Field.

In the ſame Manner I proceed round *Crabtree-Cloſe*, entering the ſeveral Occurrences as you find them in the Field-Book.

At ⊙ 17, I cauſe a Staff to be ſet up cloſe to the Fence where the Hedges join one another; to which Staff I meaſure ſtrait from ⊙ 17, cloſing *Outwood* at the Extremity of the Station-Line, which

which coincides with the Offset laid off from ⊙ 7, in the *Stockin*.

Having finifhed all the Fields on this Side the Lane, I return to ⊙ 6, obferving this general Rule, never to make a Tour greater than Neceffity requires, but always to clofe each Field as foon as poffible: So inftead of going up the *Lane* from ⊙ 2, if I had turned off into *Turfy-Leas*, and clofed firft *Turfy-Leas*, and then *Crabtree-Clofe*, &c. the Work had been done as well.

Being at ⊙ 6, I caufe a Staff to be fet up in a convenient Place, on the farther Side of *Garrot-Field*, as at ⊙ 18, laying the Chain through the Hedge, from ⊙ 6, towards ⊙ 18; and becaufe the Hedge belongs to the next Field, I write Hedge to *Will. Green*, the Owner of the adjacent Land.

After I obferved the Angle at ⊙ 18, I direct the Tellefcope to a Staff fet up by the River Side, and note the Degrees which the Index cuts on the Limb, *viz.* 131° 10′; and then meafure from ⊙ 18, to that Staff, 300 Links, taking Offsets on each Side the Line to the Brink of the River, as you fee in the Figure thereof; and this will be found very ufeful in all Manner of Practice, where the Bounders are very irregular, that as much Work may be performed at once fetting down the Inftrument as poffible.

From ⊙ 18, I proceed with the Chain to ⊙ 19, and from thence I meafure along the Hedge-fide that reaches from the *River* to the *Lane*; and when I come againft the Hedge that parts *Magg-meadow* from *Cow-pafture*, I write *a g.* 50 Links, being fo far diftant from the Chain-Line; and becaufe the Hedge from this Place belonged to *Garrott-Field*, I entred it fo in the Field-Book, but now it belongs to *Cow-pafture*; therefore I write *Hedge* to *Cow-pafture*.

Then

Then returning to ☉ 19, I direct the Telescope
first to ☉ 18, and then to ☉ 20, and find ☉ 19 to be
in a strait Line with ☉ 18 and ☉ 20; therefore I
enter in the Field-Book 180, or the Station-Line
continued, proceeding to obferve and enter
down the several Occurrences at ☉ 20, ☉ 21, and
☉ 22, round *Magg-meadow*, and then return to
☉ 21.

From ☉ 21, I go to ☉ 23 in *Cow-pafture*, clofing
it on the Corner, near ☉ 1 in *Charlton*-Field; and
fo is the whole finifhed, as far as relates to the
Field-work.

If there be feveral Pieces of Land belonging to
the fame Manor, *&c.* you are now furveying, that
lie difperfed in feveral Furlongs in Common
Fields adjacent to the fame Manor; you may
from one of your Stations on the Outfide of the
Plot take the Bearings to each Piece, by caufing a
Mark to be fet up thereon, and meafuring the
Diftance from that Station to each Mark : So may
each Piece be plotted in its true Form, and laid in
the fame Situation in the Plot as on the Land it
felf : And in your Table of References or Terrier,
you may infert the Name of the Furlong where
each Piece lies, with the Name of other Perfons
Land that lies round it, as a Direction to the
Steward or other Perfon, to find each Piece.

Obfer-

Observations and Dimensions of Land lying in the Parish of W——, *in the County of* L—— *Part of the Estate of* —— 31*st of* March, 1724.

Remarks	Offsets	Station Lines	Offsets	Remarks
		⊙ 1 *in Charlton-Field*		
	B	356°.10'		
		20	ag. 15	*Corner of Turfy Leas, Hedge to Turfy Leas.*
Corner of Cow Pasture, Hedge to Pasture.	ag. 80	40		
	10	280		
		300	20	
	18	563		
		⊙ 2 *in the Lane*		
338°.00'	B	338°.00'		
180.00	<	161.50		
518.00	10	0		
356.10		20	ag. 17	*Hedge to Home-Close.*
161.50		446		
		⊙ 3 *in the Lane*		
	B	1°.30'		
	>	203.30		
	20	0	10	
1°.30'		41	ag. 10	*Orchard-Hedge*
180 00 *Gate into*		204	ag. 15	*Orchard-Pales*
181.30 *Cow Pasture*	35	261	20	*Gate into the Yard*
360.00				
541.30 *Hedge to*		290	18	*Corner of Barn*
338.00 *Ditto.*	20	388	ag. 24	*Calves Croft-Hedge*
203.30		435		

⊙ 4

Remarks	Offsets	Station Lines	Offsets	Remarks
		☉ 4 *in the Lane*		
	B	349°.30′		
	⋎	168.00		
	ag. 20	90		
	13	140		
		220	0	on Stockin
Gate into Garrot	16	500		Hedge.
Field, Hedge to		626		
Field.				
		☉ 5 *in the Lane*		
	B	13°,50′		
	<	204,20		
	10	0	20	
		64	12	
	35	152		
	30	236	10	
		☉ 6 *in the Lane*		
	B	93° 30′		
	<	259 40		
Corner of Stockin	0	10 *int.*		
		into Stockin		
	6	270		
	3	500		
	40	750		
		☉ 7 *in Stockin*		
	B	193°.30′		
	<	280.00		
Hedge to Wood	23	0		
	60	335		
	45	620		
	ag. 20	668		
		680		

Remarks	Offsets	Station Lines	Offsets	Remarks
		⊙ 8 *in Stockin*		
	B	229°00′		
	<	215.30		
Hedge to Stockin	10	0		
	10	268		
		⊙ 9 *in Stockin*		
	B	268°30′		
	<	219.20		
Corner.	30	22		
Hedge to Stockin				
		300	20	*Clofe Stockin here*
Hedge to Lane, in corner next to Calves Croft.				
		⊙ 9 *ret.*		
	B	192°00′		
	>	143.00		
		20 *int.*		
		Into home Clofe.		
		90	20	*Calves croft Hedge*
		220	20	*Garden Pales.*
		350	20	*Corner of Or-*
		361		*chard Hedge.*
		⊙ 10 *in Home-Clofe*		
	B	264°40′		
	<	252.40		
		205	0	*on Orchard Hedge*
		250	20	*clofe here*
		255		
Out-lines of Orchard, Garden, &c. clofe on Hedge next the Lane.				

⊙ 8

Remarks	Offsets	Station Lines	Offsets	Remarks
		☉ 8 *ret.*		
	B	105°00′		
	<	91.30		
		19 *int.*		
		into home Cloſe.		
Hedge to Wood.	corner 5	25		
	37	353		
	28	465		
		☉ 11 *in Home Cloſe.*		
	B	193°30′		
	<	268:30		
	60	130		
Hedge to home Cloſe	ag: 61	476		
Corner	18	727		
		☉ 12 *in Home Cloſe.*		
	B	259°00′		
	<	246:20		
	ag. 28	65		
	ag. 48	268		
		667	cloſe here	

Cloſe Home Cloſe on Hedge to Lane, next Turfy-leas.

☉ 12

Remarks	Offsets	Station Lines	Offsets	Remarks
		☉ 12 *ret.*		
	B	189°,20′		
	⋁	176,00		
		20 *int.*		
		into Crabtree cloſe.		
		60	55	*Corner Hedge to Turfy-leaſe.*
		612	12	*Corner.*
		618 *int.*		
		into Charlton Field.		
		642		
		☉ 13 *in Charlton Field.*		
	B	262,40		
	⋖	253,20		
		10	28	
		510	20	*cloſe here.*
		530 *reaches firſt Station.*		
Turfy-leas, Cloſes on Corner of Hedge next the Lane.				
		☉ 13 *ret.*		
	B	103°,00		
	⋖	94,00		
Touch on Hedge to Crabtree-cloſe.	○	135		
	17	310		
		☉ 14 *in Charlton-Field.*		
	B	69°,00′		
	⋖	146,00		
	3	320		
	26	545		
		618		

Remarks	Offsets	Station Lines	Offsets	Remarks
		☉ 15, *in Charlton Field.*		
	B	5°,00		
	<	116,00		
	15	60		
		166 *int.*		
		Into Charlton Common.		
	120	337		
Hedge to	16	563		
Crabtree clofe.		645		
		☉ 16 *in Charlton-common.*		
	B	276°,35′		
	<	91,35		
		16 *int.*		
		Into Crabtree-clofe		
			22	30	*Hedge to Wood.*
		710	*clofe here.*	
Clofe Crabtree Clofe on Corner next Home-Clofe and Wood.				
		☉ 16, *ret.*		
	B	10°,20′		
	>	185,20		
	22	220		
	60	386		
	69	434		
A Gate into	50	611		
the Wood.	10	930		
	32	1110		
	86	1268		
	142	1353		
		1553		

Remarks	Offsets	Station Lines	Offsets	Remarks
		☉ 17, *in Charlton Field.*		
	B	264°,30′		
	<	74,10		
	128	225		
		318 *int.*		
		Into Clemenson's Land		
	56	370		
	25	504		
	12	784		
		1240		
Outwood Closes on Corner of Stockin, next to Clemson's Field.				
		☉ 6, *ret. into Lane.*		
	B	270°,50′		
	<	77,00		
		28 *int.*		
		33	10	Hedge to W.
		288	40	Green
		560	10	
		932	98	corner to River.
		☉ 18 *in Garrot Field*		
	B	186°,00		
	<	95,10		*Angle to a Bend in the*
	>	131,10		*River, from* ☉ 18.
		26	75	
	56	175	82	
	25	248	53	
		300		
		225		○ *Touch the Rivers Brink*
		422	145	
		536	110	
		620	116	

☉ 19,

Remarks	Offsets	Station Lines	Offsets	Remarks
		☉ 19, *in Garrot Field.*		
B		96°,30′		
<		90,30		
		0	8	Hedge to Garrot Field continu'd to River.
		180	0	
		390	34	
		558	50	ag. Hedge to Cow-pasture.
		890	76	
		1024	50	*close here.*

Close Garrot-Field on Hedge next to Lane.

Remarks	Offsets	Station Lines	Offsets	Remarks
		☉ 19 *ret.*		
B		186°,00′		
<		180,00		
		8 *int.*		

Into Magg-Meadow.

Remarks	Offsets	Station Lines	Offsets	Remarks
		57	126	
		143	120	
		280	42	
		348	21	
		572	97	
		665	46	
		780	8	A Bridge.
		900	0	
		1004	25	
		1045		

☉ 20,

Remarks	Offsets	Station Lines	Offsets	Remarks
		☉ 20, *in* Magg-Meadow.		
	B	151°,00′		
<		145,40		
		78	40	
		154	82	
		280	75	
		395	30	*Corner to River.*
		☉ 21, *in* Magg-Meadow.		
	B	54°,30′		
<		83,35		
		0	28	*Hedge to Magg Meadow.*
		100	64	
		245	78	
		380	59	
		452	27	
		☉ 22, *in* Magg-Meadow.		
	B	358°,00′		
<		123,20		
		0	28	
		147	53	
		378	28	*A Gate.*
		600	6	
		790	30	
		890	60	
		1010	137	*Clofe here.*
		1032		

Magg-Meadow clofes on the corner of the Hedge next Garrot Field.

Remarks	Offsets	Station Lines	Offsets	Remarks
		ʘ 21, *ret.*		
	B	143°,35′		
	<	172,35		
		22 *int.*		
		Into Cow-pasture		
		90	6	
		244	60	*Corner to River.*
		ʘ 23, *in Cow-pasture.*		
	B	92°,50′		
	<	129,05		
		205	55	*Hedge to Cow*
		245	60	*pasture.*
		302	68	
		428	24	
		560	20	
		680	38	
		755	58	
		842	45	*Close here.*
		936		
Cow-pasture closes on Corner next Charlton-Field.				

SECT. II.

*The manner of Protracting the Observations contained
in the preceding Field-Book.*

THE Protractor for this Purpose is best
made a whole Circle, and marked on the
Limb where the Numbers begin with *N.* or a
Flower-de-luce, for then may that Part of the Protractor be kept always one Way as the Instrument
in the Field, and therefore you will be less liable
to mistake, than if you use a Semicircle, which
must be laid upwards or downwards, as the
Degrees of the Bearing are more or less than 180;
and the Diameter of this Protractor is laid
Parallel to the Meridians, by the Help of equal
Divisions graved on the Protractor.

Being provided with a Sheet of strong
Cartridge-Paper, or (if that is not large enough) a
Skin of Parchment; or which I reckon better
(especially for the fair Plott) if one Sheet of
Paper be not big enough, to have several Sheets
pasted on Cloth or Canvas well stretched and
dried in a Frame before you use it; and this you
may have of any Size, as the Largeness of the
Work to be laid down requires.

Or a Practiser may have Sheets of large Paper
printed from a Copper-plate, with fine Meridian-
Lines drawn thereon at exact Distances, and
these will be very neat and true, and will save
much Trouble in drawing Meridian-Lines by a
Parallel Ruler, otherwise.

If you have not a Parallel Ruler, you may
draw Lines parallel to one another, by setting
one Foot of a Pair of Compasses at or near the
End of your given right Line, and with the other
describe the Arch of a Circle; do the same at the
other

other End of the Line, and through the utmoſt Convex of theſe Two Arches you may draw a Line Parallel to the firſt.

Having drawn Parallel Right Lines at convenient Diſtances throughout the Paper marked with N. S. repreſenting Meridian, or North and South Lines, I pick out ſome Place in one of theſe Lines, to repreſent the firſt Station, as at ⊙ 1, Fig. 22, and lay the Center of the Protractor on the Point ⊙ 1, the Diameter being Parallel to the Meridian Line, and the Beginning of the Degrees of the Protractor towards N. or upwards; and becauſe the Bearing of the firſt Station Line is 356°, 10′, I make a Mark with my Protracting Pin againſt that Number, cloſe to the Limb of the Protractor, to which Mark I draw an obſcure Line from ⊙ 1, repreſenting the Chain Line from ⊙ 1, to ⊙ 2.

Then the Field-Book being open before me, I lay the edge of my Plotting-Scale to this obſcure Line ⊙ 1 and ⊙ 2; and becauſe I find in the Field-book that the Offsets from this Line were laid off at 20, 40, 280, 300, and 563; therefore making the Beginning of the Numbers on the Plotting-Scale to coincide with ⊙ 1, I make a Prick againſt each of theſe Numbers, cloſe to the Edge of the Plotting-Scale; and then turning the Scale perpendicular to the Line, I apply it ſucceſſively to thoſe ſeveral Points, and there prick off the Length of the ſeveral Offsets on the reſpective Sides of the obſcure Line; ſo againſt the firſt Mark in the obſcure Line, I prick off 15 Links to the Right, which gives the Corner of *Turfy-Leas*: Alſo againſt the ſecond Prick in the obſcure Line, I prick off 80 Links to the Left, which give the Corner of *Cow-paſture*; at 280, or the Third Mark in the obſcure Line, I prick off

10 to the Left; at 300, 20 to the Right; and at the
End of the Line 563, I prick off 18 Links to the
Left: Laftly, I draw Lines with Ink from Point to
Point on the Outfides of this obfcure Line, thereby
conftituting the Bounders of the *Lane* fo far.

As the firft and fecond Diftances, I was againft
the Corners of *Turfy-Leas* and *Cow-pafture*;
therefore, with a Black-lead Pencil I draw Two
fhort Lines, cutting the *Lane*, to denote that the
South Fences come up to the *Lane* at thofe
Corners, and will hereafter be of Ufe in clofing
thefe Plots.

Having thus finifhed my firft Length, I produce the obfcure Line, if Occafion requires, both
Ways, till it is as long each Way as the Radius of
the Protraƈtor; then I place the Center of the
Protraƈtor on the Point ʘ 2, and turn it about
thereon, keeping the Beginning of the Degrees
towards ʘ 1; the laft Station; till the Diameter
coincides with the Station-Line ʘ 1, ʘ 2; then
clofe to the Edge of the Protraƈtor, right againft
161° 50′, the Degrees of the prefent Angle, I
make a Mark with my Protraƈting-Pin; and to
that Mark from ʘ 2, draw an obfcure Line, reprefenting the Station-Line, from ʘ 2, to ʘ 3.

And that I may be fure the Line ʘ 2, ʘ 3, is
drawn in its true Pofition, I turn about the Protraƈtor, the Center ftill coinciding with ʘ 2, till
the Diameter be Parallel with the Meridians; the
Beginning of the Numbers of the Protraƈtor
being towards N. on the Meridian Line, and then
will the Line ʘ 2, ʘ 3, before drawn, meet the
Limb of the Protraƈtor againft 338°, 00′, the
bearing of the Line ʘ 2, ʘ 3, which proves the
Line ʘ 2, ʘ 3, to be truly laid down.

And thus may the Plott be laid down by the
Angles, and examined by the Bearings.

The

The constant Rule I observe in drawing the Angles is this: To lay the Diameter of the Protractor on that Line which brought me to the present Station, where the Angle about to be laid down, was taken; and to keep the Beginning of the Numbers on the Protractor towards the last Station.

And in Order to prove that each Angle is truly laid down, I turn about the Center of the Protractor on the Point, representing the present Station, till the Diameter be parallel to the Meridians, with the Beginning of the Numbers towards N. on the Meridian-Line; then will the Line last drawn, cut the Number, expressing its Bearing on the Limb of the Protractor, if that Line be drawn in its true Position.

In like Manner, I lay down and prove the Angles taken at the 3d, 4th, 5th, 6th, 7th, 8th and 9th Stations; and also the corresponding Lengths and Occurrences, continuing the Bounders to the several Offsets as I go along, drawing a short Line across them with a Black-lead Pencil, where the Remarks, *a*, *g*, *&c.* are noted in the Field-book, breaking off the Fences where there are Gates: So at the last Length from ⊙ 9, when I have drawn that Line in its true Position, and made it of its just Length, as noted in the Field-book, I find its Extremity to coincide with that Point in the Fence, to which I laid off an Offset from 388 in the Station-Line ⊙ 3, ⊙ 4, in the *Lane*, which proves that the Angles and Lengths enclosing the *Stockin*, are truly laid down.

But if the Extremity of the last Line does not coincide with the Extremity of the last Offset laid off from the Station-Line ⊙ 3, ⊙ 4, both denoting the South-west Corner of the *Stockin*, the Lines and Angles designed to enclose the same

Stockin

Stockin, are not truly laid down, and therefore muſt be corrected before I proceed.

Next I lay the Protractor on ☉ 9, the Diameter coincident with the Line ☉ 8, ☉ 9, being the ſame Line which brought me to the that Station; and having laid down the Angle, ☉ 8, ☉ 9, ☉ 10, I prick off the ſeveral Offsets, marking them as the Field-book directs, where the Lines of Partition within come up to the Hedge.

Having drawn the Chain-Line from ☉ 10, and ſet off the laſt Offset therefrom 20 Links, I find the Extremity of that Offset to coincide with the Mark I drew croſs the Fence at the 2d Offset from ☉ 3, which proves the Work to be truly laid down.

The Out-Lines round the Orchard, Garden, Yard, &c. being drawn, the Angles within, about the Buildings, may be meaſured with a Bevel; or elſe with the Chain only, (as directed in the Uſe of the Chain) which, together with the Remarks on the Outſides, will be an eaſy Direction for drawing the ſeveral Bounders within thoſe Lines.

Having finiſhed the Ground-plot about the Buildings, I find the next Station in my Field-book, marked ☉ 8, *ret.* therefore I return to ☉ 8, in my Draught, and lay down that Angle by the Line immediately preceding that Station, *viz.* ☉ 7, ☉ 8, and proceed to ☉ 11, laying down the ſeveral Occurrences as noted in the Field-book.

The Rules I obſerve in theſe Caſes, are, to number with a Black-lead all the Stations I have already laid down in my Draught, and to expreſs thoſe Numbers ſucceſſively one after another, in a Piece of a waſte Paper.

If the Number of the preſent Station doth not immediately ſucceed that of the laſt, but is

greater

greater by an Unit than any of the Number in the
waste Paper, then I lay down the Angle of the
present Station with the Line I measured immediately before I came to it, and number it as in
the Field-Book.

But if the Number of the present Station is
greater than any in the waste Paper by more than
an Unit, there hath been some Omission in the
waste Paper, which must be rectified.

If I come to a Station whose Number is
already entred in the waste Paper, then I return
to that Station in my Draught, and there lay
down that Angle with the Line measured,
immediately before I came to this Station the
first time.

Thus observing these Directions, may the
Plot be laid down without any Burthen at all to
the Memory; and if it was surveyed by one Man,
it may be plotted by another, provided the
Person who surveyed it did observe these Rules;
and any Method of keeping a Field-book, that
lays a Burthen on the Memory, is imperfect, and
not fit for Practice.

The remaining Part of the Work is reserved
for the Exercise of the Reader: The Plan thereof
was here laid down by a Scale of $\frac{1}{4}$ of an Inch: I
sometimes lay down the Plot of each Field by a
Scale of half an Inch or larger, if the Plot will lie
on one Sheet of Paper, and cast up the Content of
each separately by that Scale; and for this
Purpose, the Sheets with Meridians ready
printed thereon, are very serviceable, though I
afterwards lay down the whole Plan together by
a Scale of a Quarter of an Inch or less, entring the
Content of each Field, as cast up by the large
Scale in the Middle thereof.

S E C T.

SECT. III.

Of Reducing Plots.

THE Plot of a Manor or Lordſhip conſiſting of ſeveral Hundred Acres lying together, being laid down by a Scale of a Quarter of an Inch or leſs, may yet be larger than is deſired, and therefore muſt be reduced into a leſſer Compaſs. Now for the Performance of this Work there are ſeveral Inſtruments, as, a long Scale made with a Center-hole at One Third Part thereof; ſo that Two Third Parts may be numbred the other Way with the ſame Number of equal Parts, though leſs according to what Proportion you pleaſe; but to paſs by this, and ſeveral others, I ſhall only give an Inſtance of the Parallelogram, which for Generality, Exactneſs and Diſpatch ſurpaſſeth all others, and is commonly made of Six Wooden Rulers joined together, and ſupported by Braſs Feet, with Holes in the Rulers for ſetting the Inſtrument to certain Proportions.

But I have lately ſeen one of theſe Inſtruments made of Braſs in a different Form from the other, and much better, becauſe it may be ſet to any given Proportion whatever, by the Help of Sliding Centers, that are moved along certain Lines calculated for that Purpoſe, and divided on the Sides of the Parallelogram; ſo that a Plott may be reduced with the utmoſt Exactneſs to any given Ratio, in Reſpect of the former, either in Proportion, as the Length of the Sides of the foul Plott ſhall be to the fair one, or elſe as the Area of the one to the Area of the other; and another Thing may be ſaid of this Inſtrument, that Curves are as well reduced thereby as Right-Lines;

Right-Lines; which by any other Inftrument is exceeding difficult, if not impoffible to be done.

The Parallelogram being fixed upon a very fmooth and even Table, and the foul Plott, and fair Paper faftened thereon, one over-againft the other, fet the Parallelogram to what Proportion you would have your reduced Plott be of, in Re- fpect to the former; then bring the Point of the Tracer to one of the outmoft Angles of the foul Plott, and put in the Point which is to draw, in its Place, letting it reft on the fair Paper; then move the Tracer with a gentle, equal Motion, over all the Lines of the foul Plott; fo fhall the Motion thereof occafion the Drawing-Point to draw upon the clean Paper or Parchment, the true and exact Figure of the former Plott, tho' of another Bignefs, according to what Proportion you fet your Inftrument; which will better ap- pear by feeing the Inftrument once ufed, then Words can poffibly explain.

When you have gone round the Lines that en- clofe one Field, you may take out the Drawing- Point, and bring the Tracer to any other Point on the foul Plott; then put the Drawing-Point in its Place again, and proceed on with your Work.

Note, This Inftrument is ufeful, not only for this Purpofe, but alfo for Copying any fmall Print, &c. in Miniature. But for Reducing great Plotts of Land, it fhould be made of a larger Size than is commonly ufed for other Purpofes.

SECT IV.

SECT IV.

Directions for Beautifying and Adorning of Plotts.

HAving reduced the Plan of the Lordſhip, Manor, *&c.* to the intended Bigneſs; 'tis neceſſary to draw imaginary Lines both vertical and horizontal, denoted by Letters at the Top and Bottom, and alſo on the Sides, to be referred to by the Table of References, for the ready finding of any Field or Parcel of Land therein contained, ſuch as you will find in the new Maps of *London*, *&c.*

The *North* Part of the Plott is always ſuppoſed to be placed upwards, and the *Eaſt* to be on the Right Hand.

The Repreſentation of Hedges ought to be laid down on the ſame Side of the Fences that they are on the Land, and to be broke off where there are to be Repreſentations of Gates.

The Out-Borders of the Plott, at leaſt ſuch as border next to the Demeſnes, ought to be fill'd with the adjacent Hedges, and the Tenants or Owners Names of the Grounds.

If you deſcribe all Rivers, Highways, Windmills, great 'lone Trees, Gates, Stiles, *&c.* that fall within your Plott, it will add to the Beauty thereof.

The Ground-Plott of Buildings, ought in all Caſes to be expreſſed by the ſame Scale that the reſt of the Plott was laid down by, and to be taken Notice of in the Table of References; but never go about to draw the Repreſentation of an Houſe or Barn in the Midſt of the Plott, ſo big as will cover an Acre or two of Land.

But if you would expreſs a Gentleman's Seat,

or

or Manor-houfe, 'tis beft done in fome Corner of the Draught, or in the Plan by itfelf, annexed to that of the Eftate to which it belongs. And the Houfe muft be drawn in Perfpective, (as you will be fhewed hereafter) and if the Gardens, Walks, and Avenues to the Houfe are expreffed, it muft be in the fame Manner; and where there are Trees, they muft be fhadowed on the light Side.

If you will take the Pains, you may, in one of the upper Corners of the Plan, draw the Manfion-houfe, &c. in the other the Lord's Coat of Arms, with Mantle, Helm, Creft and Supporters, or in a Compartiment, blazoning the Coat in its true Colours: In one of the Corners at the Bottom, you may defcribe a Circle, with the 32 Points of the Mariners Compafs, according to the Situation of the Ground, with a *Flower-de-luce* at the *North* Part thereof, ever allowing the Variation of the Needle: And in the other Corner, make a Scale equal to that by which the Plott was laid down, adorning it with Compaffes, Squares, Ovals, &c.

Having wrote the Name and Content of each Clofe about the Middle thereof, you may, about the Bounds of each Field or Enclofure, with a fmall Pencil and fome tranfparent Colour, neatly go over the black Lines; fo fhall you have a tranfparent Stroke or Margin on either Side of your black Lines, which being fhadowed, will add a great Luftre and beauty to the Plott.

If you would have your Fields all coloured, it will not be amifs to pounce over the Paper or Parchment with fome Stanifh-Grain and burnt Allom, and a double quantity of Rofin, finely fearced and lightly pumiced, to preferve the Paper from being pierced through with the Colours; or wet it all over with Allom-Water, which

which will add to the Luftre of the Colours.

Then lay on the Colours in Manner following, being firſt ground, and bound with Gum-Water very thin and bodileſs: Arable for Corn, you may waſh with pale Straw-Colour, made of yellow Oker and White-lead; for Meadows, take Pink and Verdigreaſe in a light green; Paſture in a deep green of Pink, Azure, and Smalts; Fenns, a deep green; as alſo Heaths of yellow and Indico; Trees, a ſadder green, of White-lead and Verdigreaſe; for Mudd-Walls and Ways, mix White-lead and Ruſt of Iron, or with Okers brown of *Spain*; for white Stone, take Umber and White, Water or Glaſs may be ſhewn with Indico and Azure or Black-lead; for Seas, a greeniſh Sky-Colour, of Indico, Azure, Smalts, White-lead and Verdigreaſe.

Having waſhed your Pencil very clean, take a ſmall Quantity of the Colour, and on the Inſide of the bounding Line draw the Colour along, of equal Breadth, as near as you can, broader or narrower as the Field in Bigneſs; and having gone around the Field in this Manner, ſwill the Pencil in fair Water, and ſtrike along the Inſide of the coloured Line, bringing it more down towards the Middle of the Field; and this will ſoften your Colour, and make it ſhew as if it loſt itſelf by Degrees to the very Colour of the Paper: Laſtly, with a Pen take ſome of the Colour which ſhadoweth the Colour you laid on the Field, and go over the Black-lead Line only; ſo ſhall your Field be finiſhed.

In this Manner, you may make 100 Fields on one Plott of divers Colors, obſerving as near as you can, not to colour Two Fields adjoining to one another of the ſame Color; and therefore it will be convenient to underſtand what Colors

beſt

beft fet off one another; and as near as you can, lay the Enclofures adjoining to one another, of Two fuch Colours, that one Shadow may ferve both.

This Colouring and Adorning of Plotts, is rather the Painter's Work than the Surveyor's: Yet if he has Time to fpare, and Patience to finifh the Work, it may prove a pretty Diverfion: But for general Practice, I would recommend the Ufe of *Indian*-Ink, which ground very fine upon a fmooth Tile, and the Hedges or Bounders of each Field fhadowed therewith, will look very neat, and make a Plott refemble one done from a Copper-Plate, if rightly managed.

The Water-Colors before mention'd, you may have in Shells ready prepared (being much readier than to trouble yourfelf with Grinding, &c.) as alfo the *Indian*-Ink, Pencils, &c. at Mr. *Keyton's*, a Colour-Shop, in *Long-Acre*, *London*.

C H A P.

CHAP. V.

Shewing how to meafure any Piece of Land, by the Chain only.

H E Content of any Piece of Land may be found, or a Plott thereof made, by the Chain only: So that if a Perfon is not furnifhed with Inftruments more artificial, he may work with the Chain only in the following Manner, though it be fomewhat laborious and tedious.

SECT. I.

Let Fig. 23. *be the Reprefentation of a Field, whofe Content in Acres in defired, without any Plott thereof.*

First, I walk about it, and fet up the Marks at the feveral Angles, *a, b, d, e, g, h, k, l, n,* viewing (as I go along) from which Angle to which Angle it will be moft convenient to run a Diagonal or Bafe Line, as the Line *a, g*; fo that a Perpendicular from the oppofite Angles, as *b* and *n,* may fall upon this Diagonal or Bafe, in a convenient Manner at Right Angles: And *note,* we commonly chufe the longeft Line between any Two oppofite Angles that form the Trapezia or Triangle, to be the Bafe Line.

Having

Having a Sheet of Paper in Readineſs, on which to draw an Eye-Draught of the whole Work, I make ſome Remark near the Angle, at *a*, and lay the Chain thereto, ſtretching it in a ſtrait Line towards the oppoſite Angle *g*; then I draw a ſtrait Line on the Paper, to repreſent the Line *a, g*, which I am about to meaſure; and proceed with the Chain towards Angle *g*.

When I have meaſured 4 Chain on the Line *a, g*, I perceive my ſelf almoſt over-againſt the Angle *b*, therefore having laid the Chain a Fifth Time, I ſet down an Arrow at the End of the Chain next *g*, and let it lie on the Ground in the Direction of *a, g*, and then endeavour to find a Point in the Baſe *a, g*; from which, a Right Line meaſured in the Angle *b*, may ſtand at Right Angles upon the Baſe *a, g*.

For this purpoſe I provide a ſmall Braſs Croſs, with four plain Sights thereon, having a Socket on the Backſide, which I put on the Head of a ſhort Staff, and ſet the Staff in the Ground cloſe to the Chain; then I ſet two of the Sights in the Direction of *a, g*, by looking backwards and forwards through thoſe Sights till I can ſee the Mark in each Angle *a* and *g*; ſo ſhall the Mark in the Angle *b*, be ſeen through the other Two Sights, if the Staff be ſet in Perpendicular from *b*, but if not, I move the Staff backwards and forwards by the Side of the Chain, in the Line *a, g*, till through Two of the Sights I ſee *a* and *g*, and through the other Two the Mark at *b*; then will the Staff be ſet in the Point *c*; and this is the exacteſt Way.

But if you have not ſuch a Croſs, get a Bit of Board made exactly ſquare, in Form of a ſquare Trencher, and from Corner to Corner draw ſtrait Lines thereon; and ſtick a Pin faſt and upright in

the

the Line near every Corner, and make an Hole in
the Middle of the Board, fo that you may turn it
on the Head of a Staff; this, for once, may fupply
the Want of the Crofs.

Having found the Point *c*, in the Line *a*, *g*,
there fet down a Staff, and take Notice how many Chains and links it is diftant from *a*, *viz.* 418,
therefore I apply a Scale to the Right Line on
my Eye-Draught, and near 418 make a Mark at *c*,
but don't regard whether it be exact or no, fo it
be within 20 or 30 Links, becaufe the true
Lengths of each Line are meafured on the
Ground.

Then I meafure the Ground, in a ftrait Line,
the neareft diftance between *c* and *b* 600 Links,
and erect a Perpendicular on the Point *c* on the
Paper, as near as I can guefs, by applying the End
of the Scale to the Line *c*, *a*, and by the Edge
thereof draw the Line *c*, *b* fetting clofe thereto
600 Links, the Length of the Perpendicular; but
don't regard whether the Line *c*, *b*, on the Paper,
be made of its juft Length or not, but only that
the Form of the Eye-Draught may be fomething
like that of the Field, a Refemblance thereof
being all that is requifite in this Cafe.

Next I return to the Arrow, and meafure
forwards on the Line *a*, *g*, till I am near againft
the Angle *n*; and by a few Trials, I find the Point
o, at the Diftance of 616 Links from *a*; and then
meafure the Perpendicular *o*, *n*, fetting the true
Length thereof 368, clofe to its Reprefentative
on the Eye-Draught.

Now I look towards the Fences *a*, *b*, and *a*, *n*,
and becaufe they are ftrait Lines, I draw *b*, *a*,
a, *n*, on the Eye-Draught, fo is that Side of the
Field finifhed.

From *o*, I proceed with the Chain in a ftrait
Line

Line to *g*, and find the whole Length from *a* to *g*, 1375 Links; which I fet clofe to the Right Line on my Eye-Draught.

Then I make that Right Line on the Paper nearly the Length of 1375, and draw the Right Lines *g*, *b*, and *g*, *n*, cutting the Ends of the Perpendiculars *c*, *b*, and *o*, *n*; fo fhall the greateft part of the Field be exprefled on Paper by the Trapezia *a*, *b*, *g*, *n*.

From *g*, I meafure in a ftrait Line towards *d*, and when I come to *f*, I meafure the Perpendicular *f*, *e*, 60 Links, and then meafure out the Line *g*, *d*, to *d* 800 Links, and draw the Lines *g e*, *e d*, on the Eye-Draught, to the Extremity of the Perpendicular *f*, *e*. Now I number the Angles round the Field, if I can fee the Marks at each, and compare 'em with thofe on the Eye-Draught, and thereby difcover what Part of the Field I have already meafured, and what remains to be done.

In the fame Manner as I meafured the Triangle *g*, *d*, *e*, I meafure the Triangles *g*, *h*, *k*, and *k*, *l*, *n*, and then if there be fo many Angles exprefled on the Eye-Draught, as there are Marks at each Angle round the Field, and if the true Length of each Bafe and Perpendicular, as meafured in the Field, be exprefled on the Reprefentative of the Eye-Draught, you may proceed to caft up the Work.

The fame Lines in the Field, *Fig.* 23, are meafured on the Land with the Chain, as we fhould have done on the Paper, with the Scale, if the true Plott of the fame Field had been firft made by the plain Table or other ftanding Inftrument; and therefore muft be caft up in the fame Manner; for you have the true Length of each Bafe and Perpendicular given on the Eye-Draught; fo

the

the Sum of the Two Perpendiculars *c*, *b*, and *n*, *o* multiplied by the Bafe *a*, *g*; alfo the Bafes and Perpendiculars of the other three Triangles multiplied together, and added into one Sum, the half of that gives the Content of the Field, *Fig.* 23, in fquare Links, which reduce in Acres, &c. as directed in Chap. I.

This is the beft Way I can prefcribe, for finding the Content of a Field by the Chain only, without making a Plott thereof, and is only proper for plain, level Ground, and fmall Enclofures: But even then, if the Fences are very irregular, 'tis better to go round and meafure the Angles by the Chain, taking Offsets from the Station-Lines to the Fences.

Indeed if the Fences be tolerably regular (but there are many Fields, whofe Sides are not fo ftrait as *Fig.* 23.) you may make a true Plott thereof well enough, by obferving at how many Chains and Links the Perpendiculars meet the Bafe of the Triangles or Trapezias, by erecting Lines on thofe Points Perpendicular to the Bafe, which made of their true Lengths, as meafured on the Ground, the Sides of the Field may be drawn from the Extremities of the Bafe, to the End of each Perpendicular. Alfo obferve, if any of thefe Bafes be longer than 10 Chains, leave a Staff at the Tenth, and take the Direction of the Line by that Staff, becaufe the Station will be at too great a Diftance; and if the Length be 20, fet up another Staff, and fo on.

S E C T.

SECT. II.

How to meafure Angles by the Chain.

TO this End provide Three round Station-Staves, four or five Foot long a-piece; and alfo take Care that the Ring in the Middle of the Chain, and alfo thofe at each Tenth Link be at their due Diftance from the Chain's End.

1. In order to meafure the Angle *d, o, e, Fig.* 24, fet one of your Station-Staves as upright as you can at *o*; and putting the Ring at one End of the Chain over it, let one of your Affiftants take the other End in his Hand, and ftretch out the Chain towards *d*, whilft you ftanding at *o*, direct him to move fide-ways till the Station-Staff which he has in his Hand, be brought into one Right Line with *o, d,* as at *a,* and there let him leave the Staff.

Then let him, with the End of the Chain in his Hand, move towards *e*; and as before, direct him to plant the Third Staff upright in the Line *o, e,* at *b.*

Meafure the Diftance *a, b,* in Links and Tenth Parts, if lefs than one Chain, and enter 'em in the Field-Book 88 ½.

When you plott this Angle, take with a Pair of Compaffes from a large Scale, the Diftance of one Chain; and having drawn a Right-Line *d, o,* fet one Foot of the Compaffes in *o,* and with the other defcribe an Arch *a, c*; then from the fame Scale take 88 ½ Links; and fetting one Foot of the Compaffes at *a,* let the other fall in the Arch *a, c,* and make a Mark at *b*: Laftly, through this Mark, from *o,* draw the Line *o, e,* conftituting the Angle *d, o, e.*

Obferve to plott your Angles by a large Scale,

as an Inch, or two Inches, and the Length of the Sides by a fmaller, as a Quarter of Half an Inch: Alfo obferve, that when the Length of the Chord *a*, *b*, is longer than one Chain, then 'tis beft to lay out a Sextant, or two Sextants in the following Manner.

2. The Manner of meafuring Angles with the Chain, by laying of Sextants, is deduced from this known Property of the Circle, *viz. The Radius of every Circle is equal to the Chord of One Sixth Part* (or a Sextant) *of its Periphery.*

Let it be required to meafure an Angle *b*, *a*, *c*, *Fig.* 25: Firft, fet up a Staff at *a*, and lay the Chain ftrait in the Direction of *a*, *b*, to *i*, and at 50 Links fet down an Arrow at *o*; then let your Affiftants hold the Ends of the Chain at *o*, and *a*, whilft you with the Middle in your Hand, laying both Halves ftrait, fet down an Arrow at *e*, conftituting the equilateral triangle *o*, *a*, *e*, a Sextant.

But if you have two Chains, you may (which is better) lay out the Sextants, fo that each Side of the equilateral Triangle be one Chain.

Now the Chain's End ftill held at *a*, ftretch it through the Point *e* to *d*, where alfo fet down an Arrow: Laftly, meafure the Diftance from the Arrow at *d*, to a Staff fet up one Chain's Length from *a* at *u*; fo fhall the Diftance *d*, *u*, be 76 Links, and Four Tenths of a Link; Therefore enter in the Field-Book 1^S, 764^P, implying 1 Sextant and 764 Parts.

In order to plott this Angle, *b*, *a*, *c*, thus meafured, chufe fome Line divided into 1000 Parts, and making this Line Radius, fet one Foot of the Compaffes in *a*, and with the other defcribe the Arch, *i*, *z*; and the Compaffes continuing at the fame Extent, fet one Foot in *i*, and with the other,

other, crofs the Arch at *d*, and there make a Mark.

Then take 764 Parts from the fame Line, divided into 1000 Parts, which you made Radius; and fet one Foot of the Compaffes in the Mark at *d*, and let the other crofs the Arch at *u*, and there make a Mark: Laftly, from *a*, draw a Line through the Mark at *u*, and you will conftruct the Angle required.

If you have not a Line (which is beft) actually divided into 1000 Parts, ufe the largeft Diagonal Scale you have; fo you may take off 76 Parts exactly; and the four Tenths you muft guefs at by moving the Compaffes near half way in the Diagonal towards 77; alfo obferve, the 10th of a Link is meafured on the Land by the Offset-Staff, having a Link or two thereon, divided into Ten Parts.

3. If the Angle be more than Two Sextants, as in *Fig.* 26; then having, as before, laid off the Sextant *e*, *o*, *a*, let your Affiftants hold the Ends of the Chain at *a* and *e*, while you with the Middle in your Hand, fet down an Arrow at *x*, conftituting another Sextant *e*, *a*, *x*.

Then the Chain being held at *a*, lay it thro' *x*, and at the other End *d*, fet down an Arrow: Laftly, meafure *d*, *u*, which fuppofe to be 42 Links and 5 Tenths; therefore enter in the Field-Book 2^S, 425^P, fignifying 2 Sextants and 425 Parts.

And if you would protract the Angle of *Fig.* 26, then with the Length of the Line divided into 1000 Parts, defcribe the Arch *i*, *y*, and thereon lay *i*, *n*, and *n*, *d*, each equal to the Radius or divided Line; and afterwards lay 424 equal Parts from *d*, to *u*, and draw *a*, *u*; which gives the Angle, as required.

Obferve,

Obferve, if you were about to meafure the Angle, *Fig.* 26, and had fet up one of your three Staves where the Station-Lines meet in the angular Point *a*, another at *i*, and the other at *u*, in the Lines *a*, *b*, and *a*, *c*; before you proceed to meafure the Angle *i*, *a*, *u*, you muft be fure that the Staves at *a*, and *i*, and the Mark at *b*, are exactly in the fame Plane; and alfo the Staves at *a*, and *u*, and the Mark at *c*, in another Plane.

So when the Staff at *a*, is planted as nearly Perpendicular as you can, move yourfelf backwards, the farther the better, 'till you fee the Staff at *a*, and that at *i*, in one ftrait Line with the Mark at *b*; there ftand, and direct your Affiftant to place his Staff, fo that the Staff at *a*, exactly cover that at *i*, from the Top to the Bottom.

SECT. III.

Obfervations on Working with the Chain.

IF you would continue a ftrait Line, you may fignify it, by entring in the Field-Book 3^S. 000, that is, 3 Sextants.

If an Angle be external, and fo contain more than 3 Sextants, as *b*, *a*, *e*, *Fig.* 27, put the Ring at one End of the Chain over the Staff at *a*; and taking the other End in your Hand, ftretch out the Chain at Length towards *d*, and move Sideways, till you perceive yourfelf in a Right-Line with *a*, *b*, and there at the End of the Chain, fet down an Arrow at *d*, fo that *d*, *a*, *b*, are in the fame Plane, and then fet down the other Staff at *c*, at the End of one Chain alfo; fo that the Staves at *a*, and *c*, be in the fame Plane with the Mark at *e*. Now meafure the Angle *d*, *a*, *c*, in the fame Manner as aforefaid, and to it add the Sextants,

fo

fo will the Sum be the Meafure of the external
Angle *d*, *a*, *c*.

So if the Angle *d*, *a*, *c*, be 947, then will the
external Angle *b*, *a*, *c*, be 3 Sextants, 947 Parts;
and if the Angle *d*, *a*, *c*, be 1^S. 947, then *b*, *a*, *c*,
will be 4^S. 947, &c.

When you protract the external Angle *b*, *a*, *e*,
firſt continue the Line *b*, *a*; then from the Angle
fubtract 3 Sextants, and make the Angle *d*, *a*, *e*,
equal to the Remainder.

2. But if you go on the Outſide of a Field or
Wood, you may then work as though you were
within the Wood, by meaſuring the Angle verti-
cally oppoſite to thoſe that are internal: So in
Fig. 28, if you meaſure the Angle *o*, *a*, *u*, inſtead
of *b*, *a*, *e*, it will do your Buſineſs when you come
to protract, as well as if you had meaſured *b*, *a*, *e*,
on the Inſide; for if two Right-Lines croſs one
another, the contrary or vertical Angles are
equal. *Euclid.* 15. 1.

Angles meaſured by the Chain, may be laid
down by a Protractor made on Purpoſe, having
Sextants and Links divided thereon; and then to
be uſed in the ſame Manner as other Protractors.

The Manner of keeping the Field-Book, is in
all Reſpects the ſame as thoſe uſed in the 2d, or
4th Chapters, except that when the Angles are
meaſured by the Theodolite, you note the
Quantity of each by Degrees and Minutes: In
this Caſe, when meaſured by the Chain, you
note the Quantity by Sextants and Parts.

So if you were to meaſure the Field, *Fig.* 14,
by the Chain, inſtead of noting 102°, 20′, for the
Quantity of the Angle *b*, you muſt note 1 Sextant 734 Parts; and inſtead of 230°, 50′, for the
external Angle *d*, you muſt note 3 Sextants, 886
Parts; but the Station-Lines, Offsets, &c. will ſtill
be

be the fame.

There are other Ways of Working with the Chain; but thefe before-mention'd are the beft and exacteft, and contain as much Variety as any one will commonly put in Practice: Alfo thereby you might meafure an inacceffible Diftance, and do feveral other Things; but thefe are only for a Shift, when we have no other Inftruments: And the fame may be faid of meafuring Angles.

SECT. IV.

Obfervations on meafuring Land in Common-Fields.

WHen ploughed Lands in Common-Fields are meafured by the Chain, 'tis ufual to meafure the Length down the Ridge of the Land, and to take the Breadth at the Top of the Land, about the Middle, and at the Bottom; and adding thefe Three Numbers together, to take the Third Part of the Sum for the mean Breadth; but 'tis not advifeable to take the Breadth very near the Lands Ends, becaufe the Turning of the Plough generally makes it confiderably narrower or wider; and if in meafuring down the Land, you find the Breadth is not nearly equal, 'tis beft to meafure crofs the Land oftener, as at every 3 or 4 Chains Length, and adding the feveral Breadths together, divide that Sum by the Number of Breadths, for the equated Breadth: And for this Practice, half the Four Pole Chain is moft convenient, remembring either to fet them down as whole Chains, or to make 'em fo, when you caft up the Content.

The feveral Furlongs in Common arable Fields, may be accounted as fo many particular Enclofures, and meafured after the fame Manner,

by

by ſetting up Marks at the Extremities of the Furlong, and meaſuring the Angles by the Theodolite, as before directed; and as you paſs along the Station-Lines, you may from thence take Offsets to each Man's particular Lands; and againſt that Offset write the Name of the Owner or Tenant: And when you plott that Furlong, you may, by thoſe Directions in your Field-Book (if you will take the Pains, which is not a little) expreſs each particular Land in your Draught, with its Buttings and Boundings (but the Buttings and Boundings of Land in Common-Fields is neceſſarily expreſſed in all Caſes): An Example of this is needleſs, only it may be added, that in the Survey of a large Common-Field, 'tis ſafer to divide it in Parcels, as ſeparate Fields, keeping good Marks at the Stations, then to venture the Cloſing of the Plott, by going round it all at once, and dividing it into Parcels afterwards.

C H A P.

CHAP. VI.

The Manner of Laying-out, or Dividing Land.

SECT. I.

PROBLEM I.

F any Quantity of Acres be given, to be laid out in a Square Figure, annex to the Number of Acres given 5 Cyphers, which will turn the Acres into ſquare Links; then from the Number thus encreaſed, extract the Square Root, which ſhall be the Side of the propoſed Square.

So if you would cut out a Corn-Field one ſquare Acre, add to 1 five Cyphers, and then it will be 100000, the Root of which is 3 Chains, 16 Links, and ſomething more for the ſide of that Acre.

Problem 2. If you would lay out a given Quantity of Acres in a Parallelogram, whereof one Side is given. — Firſt, turn the Acres into ſquare Links, by adding 5 Cyphers; and divide that Number thus encreaſed by the given Side, the Quotient will be the other Side, as if 100 Acres was to be laid out in a Parallelogram, on Side whereof ſhall be 20 Chains; therefore to the 100 Acres I add 5 Cyphers, which divided by 20 Chain,

Chain, the Length of the given Side, the Quotient is 50 Chains for the Length of the other Side.

Problem 3. If you would lay out a Parallelogram that shall be 4, 5, 6, &c. Times longer than it is broad. First, turn the given Quantity of Acres into Links, as before, which Sum divide by the Number given, for the Proportion between the Length and Breadth, as 4, 5, 6, &c. the Root of the Quotient will shew the shortest Side of such a Parallelogram: As if it was required to lay out 100 Acres in a Parallelogram that shall be 5 Times as long as broad; first, to the 100 Acres add 5 Cyphers, and it makes 10000000; which Sum divide by 5, the Quotient is 2000000, the nearest Root of which is 14 Chains, 14 Links, and that shall be the shortest Side of such a Parallelogram; and by multiplying that 14 Chains, 14 Links, by 5, shews the longest Side thereof to be 70 Chains, 70 Links.

Problem 4. If you would make a Triangle that shall contain any Number of Acres, being confined to a certain Base: First, double the given number of Acres, to which annex 5 Cyphers, and divide that Sum by the Base, the Quotient will be the Length of the Perpendicular: As if the Base given, be 40 Chains, upon which I am to make a Triangle that shall contain 100 Acres, first I double the Number of Acres, and annexing 5 Cyphers thereto, I divide it by 40 Chains, the limited Base, so shall the Quotient be 50 Chains; for the Height of the Perpendicular, which I set on any Part of the Base, and from the Extremities thereof, draw the other Two Sides, which shall form the Triangle required.

So if when you are laying out a new Piece of Land of any given Content, altho' you meet in your Way with 100 Lines and Angles, yet you

may,

may, by making a Triangle to the first Station
you begin at, cut off any Quantity required.

SECT. II.

Of Dividing Lands.

EXAMPLE I.

SUppofe it was required to divide *Fig.* 29,
whofe Content is 10 *A.* 3 *R.* 22 *P.* into 2 equal
Parts, by a Line drawn parallel to *a, b.*

First, the Acres, Roods and Poles muft be reduced into fquare Links, which may be done thus:

If the Roods are $\left\{ \begin{array}{c} 1 \\ 2 \\ 3 \end{array} \right\}$ add $\left\{ \begin{array}{c} 40 \\ 80 \\ 120 \end{array} \right\}$ to the Poles,

and to the Sum annex 4 Cyphers; divide this
laft by 16, and write the Quotient Figures, if
they be 5, after the Acres. But if the Quotient

Figures are but $\left\{ \begin{array}{c} 4 \\ 3 \end{array} \right\}$ write $\left\{ \begin{array}{c} 1 \\ 2 \end{array} \right\}$ Cypher
Cyphers

after the Acres, and then the Quotient Figures;
fo will you have the fquare Links required.

Therefore 10 *A.* 3 *R.* 22 *P.* reduced into
fquare Links, will ftand thus:

$$\left\{ \begin{array}{c} 120 \\ 16 \ (1420000 \ (88750 \\ \text{Square Links} - 1088750 \end{array} \right\}$$

Having reduced the Acres, *&c.* into Square
Links, they make 1088750, the half of which is
544375, next draw a Line by guefs parallel to *a, b,*
as the Line *c, d,* and then caft up the Content of
the Figure *a, d, c, b,* which fuppofe 494375 Square
Links lefs than 544375, by 50000 Square Links,
which fhews that the Partition-Line muft be fet
forwarder from *b, a.*

Now,

Now, in order to know how much *c*, *d*, muſt be ſet forwarder, I divide the Exceſs 50000 Square Links by the Length of the Line *c*, *d*, 953 Links, and it quotes 52 Links; therefore from *c*, I ſet off 52 Links, and draw the Line *f*, *e*, parallel to *b*, *a*, and it will be ſufficiently near the Partition-Line.

This is performed by the ſecond Problem of the laſt Section; but if thoſe Parts of the Bounders *c*, *f*, *d*, *e*, be not nearly parallel, then 'tis beſt to draw a Triangle to *c*, *d*, inſtead of the Parallelogram *c*, *f*, *d*, *e*.

But if *c*, *d*, had cut off the Quantity *a*, *d*, *c*, *b*, greater than required, then the Partition-Line had been more towards *a*, *b*, whoſe Diſtance might be found as before.

Examp. 2. Suppoſe it was required to cut off from *Fig*. 30, 6 Acres towards *g*, *f*, by a Line drawn from a given Point in the Bounder *g*, *a*, at *a*.

Firſt, reduce the given Quantity, 6 Acres, into Square Links, and they will be 600000; and then draw the Line *a*, *b*, by Gueſs from the given Point *a*, and caſt up the Content of *g*, *a*, *b*, *f*, which amounts to 431680 Square Links, which is too little.

Next draw the Line *a*, *e*, from the Point *a*, forming the Triangle *a*, *b*, *e*, whoſe Content is 235600, which added to the Part *g*, *a*, *b*, *f*, amounts to 667280, which is more than the given Quantity 600000, by 67280 Square Links; therefore the Partition-Line paſſes between *e*, and *b*.

Now divide the Exceſs 67280, by 380, half the Length of the Perpendicular *a*, *c*, in Links, the Quotient is 77 Links, which ſet off from *e*, towards *b*, and draw *a*, *d*, which is the true Line of Partition.

Examp.

Examp. 3. Suppofe *Fig.* 31, was to be divided equally amongft Three Tenants, in fuch Manner that the dividing Lines may pafs through the Pond *o*, in the Middle of the Field, fo that each Tenant may have the Benefit of the Water.

Firft, Reduce the whole Figure into Square Links, and it will be found to contain 1477410; then each Tenant muft have One Third Part thereof, *viz.* 492470 Square Links.

From *o*, to any two Angles, as *a*, and *b*, draw the Lines *o a*, *o b*, forming the Triangle *a*, *o*, *b*; which being caft up, amounts to 291984 Square Links, which is too little.

To the next Angle *f*, draw *o*, *f*, forming the Triangle *a*, *o*, *f*, which being caft up, amounts to 231000 Square Links, which added to the Triangle *a*, *o*, *b*, gives 522984, which exceeds the Quantity required by 30515 Square Links.

Divide the Excefs 30514, by 347, half the Length of the Perpendicular *o*, *g*, and lay the Quotient 87, from *f*, to *h*, and fo fhall *h*, *o*, *b*, *a*, *g*, be One Third Part of *a*, *b*, *c*, *d*, *e*, *f*.

Next draw the Line *o*, *e*, to the next Angle *e*, and caft up the Content of *o*, *e*, *f*, amounting to 256410 Square Links; to which add the Triangle *h*, *o*, *f*, 30514 Square Links, the Sum is 286924, which is too little.

Therefore draw *o*, *d*, to the next Angle *d*, and caft up the Content of *o*, *e*, *d*, 265500 Square Links, to which add *h*, *o*, *e*, *f*, 286924, their Sum is 552424 Square Links; which is more than the Third Part of *a*, *b*, *c*, *d*, *e*, *f*, by 59954 Square Links.

Divide the Excefs 59954, by 295, half the Length of the Perpendicular *o*, *i*, and lay the Quotient 203 Links from *d*, to *k*, and draw *o*, *k*; fo fhall *Fig.* 31, be divided into 3 equal Parts, by the
Lines

Lines *b, a, h, o,* and *h, f, e, k, o,* and *k, d, c, b, o,* as was required; and the Pond *o,* laid out to each Tenant apart.

These 3 Examples express all the Variety that moſt commonly comes in Practice; for either the Partition-Line is required to be Parallel to ſome other Lines aſſigned; or to paſs through ſome given Point in the Fence; or to paſs thro' a Point aſſigned in the Land.

If a Piece of Common was to be divided a-mongſt ſeveral Tenants, in Proportion to the Rent which each pays for his Farm: The Num-bers reduced to the loweſt Denomination (except you expreſs the Parts of Acres and Pounds by Decimals, which is better) the Rule is:

As the Sum total of all Tenants Rent, is to the whole Number of Acres in the Piece of Land contained; ſo is each particular Tenant's Rent, to the Number of Acres to be laid out for his Part: This is very plain, and needs no Example.

So if a Piece of Common was to be encloſed, and divided amongſt ſeveral Tenants, according to the Number of Beaſt-Gates which each Ten-ant hath in the Common, it is to be performed (*mutatis mutandis*) by the ſame Rule.

There is no need of Direction how to make the Lines on the Land in the ſame Poſition as on the Paper-Plot, by carrying the Chain in a ſtrait Line from Point to Point, on the Land it ſelf, as divided on the Paper: Only take Notice, that the larger the Scale is, by which the Plott is laid down on the Paper, the exacter will the odd Links of each Line be eſtimated by the Scale, in order to transfer thoſe Lines to the Land.

But if you are to divide a Wood, or very hilly Ground, ſo that you can't ſee the Marks from Side to Side, do thus:

Be

Be fure to keep good Marks at every Station, as you meafure round it, that you may find the Hole at each, in which the Staff ftood; then having plotted the Wood, and divided it on the Paper-Plott, in fuch Manner as defired, plant the Center of the Theodolite directly over that Point in the Station-Line on the Land where the dividing Line cuts it, on the Paper-Plott, and bring the Index to 360, or fet it in the fame Pofition as it was at the forward Station when you meafured that Angle, turning about the Inftrument, till the Hair in the Tellefcope cuts the laft Mark; fo that the Tellefcope be exactly in the Direction of the prefent Station-Line, where the Dividing-Line cuts, and there fcrew the Inftrument faft; then meafure with your Protractor on the Paper-Plott, the Angle which the Dividing-Line makes with the prefent Station-Line; and turn about the Index on the Limb to the fame Angle; fo fhall the Tellefcope be fet in the Direction of the Dividing-Line; then by looking through the Tellefcope, you may caufe Staves to be fet up in the fame Direction: And thus proceed in a ftrait Line, till you are far enough in the Wood, or quite through, if it be divided by one Line; but if by two Lines, you muft continue them till they meet one another, as in the Paper-Plott.

The fame Thing may be performed by the plain Table, or the Chain only; but thofe Inftruments are not fo convenient to meafure a Wood, or hilly Ground, as the Theodolite.

S E C T.

SECT. III.

How to reduce Customary into Statute Measure.

IF you would change Customary into Statue Measure, & *è contra*, the Rule is: As the Square of one sort of Measure is to the Square of the other, so is the Area of the one, to the Area of the other.

In some Parts of *England*, they account 18, in some 20, 22, &c. Feet to a Pole or Perch, and 160 such Perches to make an Acre, which is called customary Measure; whereas our true Measure of Land, by Act of Parliament, is but 160 Perches to an Acre, accounting 16 Feet and an half to the Perch.

So if a Field measured by a Perch of 18 Feet, accounting 160 such Perches to the Acre, doth contain 100 Acres, how many Acres shall the same Field contain by the Statute Perch of $16\frac{1}{2}$ Feet ? Say, As the Square of 18 Feet, (*viz.*) 324, is to 100 Acres, so is the Square of $16\frac{1}{2}$, Feet, (*viz.* 272, 25,) to $119\frac{9}{10}$ of an Acre Statute.

C H A P.

CHAP. VII.

General Observations touching the Surveying and Plotting of Roads, Rivers, &c. With short Hints how to make the Draught of a County, or ground Plott of a City, &c.

N this Seventh Chapter I have added general Directions for Measuring of Roads, *&c.* omitting particular Forms of Charts, as *fac similes*, which would take up more Room than can be spared in this small Tract; and indeed if the several Varieties that occur in these large and spacious Works were inserted, it would swell to a large Volume: But since the Surveyor's Judgment in contriving and carrying on his Work must be his best Guide; these few Observations may serve as Memorandums of the most necessary Things in Practice, which, together with other Rules before laid down in this Tract, may perhaps be sufficient Instruction for the Performance of any Thing of this Nature.

SECT.

SECT. I.

General Directions for making a Draught of the Roads lying through any County, &c.

INstruments fitteft for this Purpofe are 1. the Theodolite as before defcribed: The Angles which each Station-line on the Road makes with the Meridian, being obferved by the Limb in the fame Manner, as before defcribed in Chap. III. and the Bearings of the feveral Remarks from thence by the Needle. 2. The Wheel or Way-wifer to meafure the Length of the Lines, by driving the Wheel on the Road before you, fo fhall the Hand on the upper Part of the Inftrument fhew how many Miles, Furlongs, and Poles you go at one Time from any Station. 3. The Protractor as before defcribed: A neat diagonal Scale of Brafs and a good Pair of Compaffes, or rather a Pair of beam Compaffes, with fuch a Scale on the Beam as fhall be agreeable to the Largenefs of your Plott; for thereby you may lay down the Length of your Lines much exacter than by any other Way, by fetting one Foot of the Compaffes at one End of the Line, and moving the Socket on the Beam to one of the equal Divifions near the other End of the Line you are about to lay down, reprefenting Chains or Furlongs; and then you may bring the Point of the Compafs which ftands perpendicular on the Paper to the Parts of that equal Divifion, reprefenting Links or Poles by the Help of a fmall Screw, there being Divifions on the Edge of the Socket fliding clofe on the Beam according to *Nonus's* Projection; fo that the Links of a Chain or Poles of a Furlong are eftimated in the fame Manner as the Minutes of a Degree on the

Limb

Limb of the Theodolite; for in thefe large Plans where the Diftances of Places are determined by the Interfection of Right-lines from your Stations; thofe ftationary Diftances ought to be laid down as accurately as may be, for where a Mile is laid down in the Compafs of an Inch, a Point is confiderable.

In order therefore to make a Draught of the principal Roads that lye through any County, &c. firft begin at fome noted Market Town, or rather at the County Town, placing the Theodolite at fome remarkable Church, &c. then having a Field-book with large Margins to enter the Remarks, and the middle Column reprefenting the Station-Lines divided into three Parts, at the Head of each of which write *M.* for Miles, *F.* for Furlongs, and *P.* for Poles.

When you begin your Journey at the Top of the Field-book write the Name of the Place where you began your Work, and making ☉ 1, in the Field-book to reprefent the firft Station: Send fome Perfon forwards on the Road, with a white Flag in his Hand, as far as you can fee; and then by fome known Sign caufe him to ftand; then bring the Index to 360, on the Limb, and turn the Inftrument into the Direction of the Meridian, and there fix it; then direct the Tellefcope to the Perfon on the Road, and note the Degrees cut on the Limb for the bearing of the firft Station-Line.

Put the Hands to the Beginnings of the Numbers on the Plate, and bring the Wheel to the Station, then caufe one to drive it from the Place where your Inftrument ftood towards the Man on the Road, 'till you fee fome remarkable Object on either Side thereof; there let him ftop and direct the Tellefcope to that Object, and
note

note the Degrees which the Needle points to in the Box, and at what Diftance the Inftrument is planted from the laft Station, together with the Name of the Object to which the Tellefcope was directed.

Having entered this in the Field-Book, go on with the Wheel till you fee fomething elfe re-markable on either Side of the Road; there ftop and take a Bearing there; and in this Manner proceed till you come up to the Man at the fecond Station, obferving as you go along the Road from Station to Station: Firft, what by Lanes, or Roads you meet with in your Way, whether they be to the right Hand or to the left, and to what Places they go, and how they incline, whether forwards or backwards, or whether they be at right Angles with the Road you are meafuring, and note it down in the Field-book with two Lines thus = on the right or left Side of the Station-line; that is, if the Road or Lane be on the right Hand, then place it on the right Hand; but if the Road be on the left Hand, then place it on the left: If the Road doth incline forwards, then make it on either Side of the Lane or Road thus ⊬: If the Road or Lane incline backwards, then mark it thus ⊰: If it be at right Angles with the Road you meafure, then mark it down thus ⊒: If another Road croffes that you are upon, note it thus ∓: Likewife fet down at what Diftance from your Stations the Lanes or Roads do turn out from the Road you meafure, *viz.* at fo many Furlongs, *&c.* a Road to the Right or Left to fuch a Place.

Likewife, when you pafs over any Bridge, note it in the Field-Book, with the Diftance from the laft Station; as alfo the Name of the Water that runs under it, and from whence it hath its

Rife,

Rife, and where it doth empty it felf: So muſt you do when you paſs over any Ford or Rill.

Note down alſo, when you aſcend an Hill, and when you come to the Top thereof, and when you deſcend the ſame, and come to the Bottom thereof.

When you paſs through any Town or Village, note at how many Miles, Furlongs, and Poles you enter the ſame; and at how many Miles, &c. you leave it, and whether the Houſes be cloſe, or ſcattering, or on the Right or Left Side of the Road, or on both Sides thereof; alſo write down the right Name thereof; and if a Market-Town, take Notice on what Day the Fairs or Market is kept; and by what Officers the Town is governed.

Note down alſo the Mills that are on the Road, whether Water-mills, or Wind-mills, and the Diſtance from the your laſt Station. If there be any 'lone Churches on the Road, note them down by their Names, and whether they be Towers or Spires, with their Diſtance from your laſt Station.

In your meaſuring along the Road, if you ſee any Churches, Manſion-Houſes, Beacons, Windmills, Towns, Villages, or any other Thing remarkable, you muſt take a Bearing to each, noting down in your Field-Book, the Name of the Place, and how it does bear, and at what Diſtance from your laſt Station you took this Bearing.

Then in your meaſuring forwards, at as great a Diſtance as you can, take another Bearing to thoſe Places you took laſt, provided you may but ſee them, and note the Name of the Places, and how they bear, and at what Diſtance from your laſt Station, as before.

Obſerving theſe Directions, proceed with
your

your Work on the Road as far as you can go the firſt Day, entring the ſeveral Obſervations in as plain and fair a Manner as poſſible in the Field-Book, and then it may be convenient to protract that Day's Obſervations, before you go any further.

Therefore, on the Paper or Parchment, on which you draw the foul Draught, let there be ruled Meridian-Lines all over, exactly parallel to one another; and chuſing a proper Place in one of the Lines, to repreſent the firſt Station, draw an occult Line from thence, making ſuch an Angle with the Meridian, as you obſerved the firſt Station-Line to do, when you directed the Tellescope to the Man ſtanding in the Middle of the Road.

When you have drawn the Station-Line in its true Poſition, ſet thereon the ſeveral Diſtances from the laſt Station very exactly, at which you made any Remarks, as you find 'em noted in the Field-Book; and make a ſmall Prick at each, in the Station-Line: Then having made the Station-Line of its juſt Length, proceed to lay down the ſeveral Objects you obſerved on each Side the Road, in their true Situation; as ſuppoſe a Steeple that ſtands at a Diſtance from the Road, *viz.* a Mile or two; lay the Center of your Protractor on the Place at which you took the Bearing, thus, (at ſo many Poles, *&c.* Diſtance from ſuch a Station, ſuch a Steeple did bear from you 207°, 40′,) therefore, againſt the Degrees of Bearing make a Mark, and draw a Line at Length.

Then at the ſecond Place in the Station-Line, where you obſerved this ſame Steeple to bear from you, lay the Center of your Protractor, and againſt the Degrees of Bearing make a Mark, and likewiſe draw a Line at Length; and where this

laſt

laſt Line of Bearing doth interſect the firſt Line of Bearing, there place the Steeple, with the Body of the Church to the Eaſt Side thereof.

All Wind-mills, eminent Houſes, or other Remarks that are diſtant from the Road, you muſt protract in the ſame Manner as you did the Church, by the Bearings, and likewiſe write down the Name of each; and if you protract a Village that ſtands at a Diſtance from the Road, you muſt ſignify by Writing the ſame, that it is a Village; but that you may know Market-Towns from Villages, write the Name of the Market-Town in a different kind of Letter; and if you protract a Village that is in the Road, with Houſes ſcattering, you muſt place your Houſes ſcattering on the Right or Left Hand of the Road, as you noted them in the Field-Book.

You muſt protract the Road all along with two Lines parallel one to the other. If your Road have Hedges on both Sides, then draw your Lines black; but if your Road be open Way, then draw it with prick'd Lines; alſo you may inſert the Quality of the Ground, whether it be a Common, Moor, or arable Land.

If the Road paſs through a Wood, then make little Trees on both Sides the Road, to ſignify the ſame ſo far as the Wood goes.

If the Road paſſes over an Hill, you muſt at the Beginning where the Hill doth aſcend, ſhadow very deep, and as the Hill doth more and more aſcend, you muſt ſhadow it lighter, till you come to the Top thereof: But if the Hill makes an Angle of above 5 or 6 Degrees, or thereabouts, and the Height be above a Furlong, you muſt find the horizontal Line of that Hill, and protract that, otherwiſe a great Error may enſue.

If there be a Village or Town on the Side of
the

the Hill, you muſt ſhadow it likewiſe, ſo that the Houſes may be ſeen. If the Remarks that are at a Diſtance from the Road ſtand on a Hill, make an Hill to repreſent the ſame.

If your Road paſs by or through a Park, Foreſt or Chaſe, write down on your Road protracted, where you did enter the ſame, and where you did leave it, writing the Name thereof among the Trees.

If your Road paſs over a Ford, draw the River quite croſs the Road, to ſignify there is no Bridge, and write the Name of the Ford; but if there be a Bridge, then draw the River on both Sides the Road, 'till it touch the Parallel Lines; and write the Name both of the Bridge and of the River; likewiſe write on that Side of the Road that the Stream runs from you, and at what Place the river doth empty it ſelf; and on the other Side of the Road write from whence the Water or River hath its Riſe, if you can learn that of the Inhabitants.

All Rills you may ſignify, by drawing a Line croſs the Road; and Brooks may be ſignify'd by drawing Two Lines croſs the Road, and Rivers by more Lines, together with the Names; for all Rivers have Names, but Rills and Brooks have none.

It will likewiſe be neceſſary, that you take Notice of the Quality of the Way, whether it be ſtony or clayey or boggy, and write it down on the Road that you have protracted: And by this Means you will have your Road mighty full of Remarks, and it will ſhew very delightful.

S E C T.

SECT. II.

Containing general Directions for making the Plott of a River or Brook, by the before-mention'd Instruments.

Irst, when you come to the Mouth of the River, cause a Man to go and stand at the next Bend thereof; then plant your Theodolite at the Mouth of the River, letting your Needle hang directly over the Meridian-Line in the Box; there fix the Instrument fast, and direct your Tellescope to the Man that stands at the next Bending of the River, and note down the Angle in your Field-Book, as you did in the Road.

The cause the Man that drives the Wheel, to measure between your first Station, and the Man at the next Bending; and note that down also in your Field-Book, under Miles, Furlongs, and Poles.

Then bring your Instrument to the Man at the first Bending of the River, and cause that Man to go forwards till he finds another Bending, there let him stand; and placing your Instrument where the Man last stood, let your Needle (as before) hang directly over the Meridian-Line, and there make your Instrument fast; and direct the Tellescope to the Man that stands at the next Bending of the River, and note down the Angle in your Field-Book, as you did in the former: And thus you must proceed all along the River, to the Head thereof.

In order to take the Breadth of the River, it will be convenient to send some Body on Purpose cross the River, in a Boat, (unless a Bridge or Ferry be near,) and let him set up a Staff by the

Brink

Brink of the River, on the further Side, to which Staff take a Bearing, from the Place of your Standing, which call the firft Station; alfo let another Staff be fet up on the fame Side where you ftand, and call that the fecond Station, to which take a Bearing alfo. Now meafure in as ftrait a Line as poffible, the neareft Diftance between the 1ft and 2d Stations, and that Diftance note in the Field-Book with the Bearings.

Plant the Theodolite at the fecond Station, and take a Bearing to the fame Mark on the further Side of the River, and note that Bearing alfo in the Field-Book.

When you protract thefe Obfervations, lay the Center of the Protractor to ⊙ 1, and turn it about till the Diameter be parallel to the Meridians on the Paper, the againft the Degrees of Bearing from ⊙ 1, to the Mark on the further Side of the River; and alfo to ⊙ 2, clofe to the Limb of the Protractor make 2 Marks, through which, from ⊙ 1, draw 2 Lines at Length.

Set off the Diftance between the 2 Stations on the 2d Line, and mark it ⊙ 2, to which Mark lay the Protractor as before, and againft the Degrees of Bearing obferved at this 2d ⊙; to the Mark on the further Side of the River make a Mark, through which draw a Line at Length; then will this Line interfect the firft Line drawn at your firft Station, fo fhall the Point of Interfection fhew the Breadth of the River.

In the fame Manner, are the Diftance of the Churches, &c. from your Station on the Road, determined; and in chufing the Diftance of thefe Stations, 'tis very neceffary to obferve the Rule laid down at the Conclufion of Chap. I.

If there be a Ferry over the River, you muft draw the River to its true Breadth, and make a

prick'd

prick'd Line crofs the River, to reprefent the Paffage of the Ferry-Boat; and note on the Side of the River the Name of the Ferry.

In meafuring on by the River, obferve what Bridges you pafs by, and at what Diftance from your laft Station; alfo whether they be of Wood or Stone, and by what Name they are called; alfo take Notice of all Corn-Mills, Paper-Mills, &c. and note them in the Field-Book, in the Column of Remarks, with their Diftance from the Mouth of the River, and your laft Station.

Likewife take Notice of all the Sluices (if there be never fo many) that are on the River, and of all the Locks and Flood-gates as you pafs along, with their Names, if they have any; alfo if there be any cut Rivers from the River that you are meafuring, note where it goes out of the River, and where it comes in again, and for what End it was fo cut: Alfo where any Brook or River enters into that you are meafuring, note down the Place, and the Name of the River that comes in; and alfo take an Account of thofe Places of the River that are fordable, and note them down in your Field-Book: And in all thefe Cafes, exprefs the Diftance of each Remark from your laft Station, as alfo their Diftance from the Mouth of the River.

You muft alfo note in your Field-Book all the Towns this River doth run through, or by, with the Towns Names, and the Diftance from your laft Station and the Mouth of the River.

You muft alfo take an Account of all the Churches that are on each Side the River within your View, by taking a Bearing at them at two feveral Places, as you did on the Road; and note them down in your Field-Book, with the Dift-ance of the Place from your laft Station, where
you

you took the Bearing, to the Steeple both Times; by this Means you will come to know how far each Church is diftant from the River: The fame you muft do by all the Wind-mills, great Houfes, &c. noting their Names, and Places of Situation in the Column of Remarks in the Field-Book.

When you have thus meafured your main River, begin to meafure the feveral Branches thereof; for there are but few Rivers but have fmaller Rivers running into them, and all thofe fmall Rivers ought to be done with the like Exactnefs with the great ones.

Note, All Rivers that are navigable, every Bending of them muft be taken exactly; but for other fmall Creeks there is no great Need; for you will find fuch fmall Brooks to have a Bend at every two or three Poles, nay fometimes lefs, therefore they are to be taken thus:

Take your Sights as far as you can conveniently, till you find the Brook to have confiderable Bending; and if your Scale will permit, you may take Offsets to reprefent the fmall Turnings and Windings thereof, as in *Fig.* 22: But in meafuring a fmall Brook, if your Scale is to be a Mile or two in an Inch, then thefe fmall Turnings and Windings can't be defcribed in the Map.

The Manner of protracting thefe Obfervations, is the fame with the Roads, except the Offsets from the Station-Line to the Brink of the River, and its Breadth, which are particularly to be regarded.

SECT.

SECT. III.

General Directions for making a Map of a County,
&c.

Firſt, from the County-Town, or other Market-Town, where you began your Work, lay
down the principal Roads throughout the
County, protracting them truly, as you obſerved
them in your Survey, inſerting the Towns, Villages, great Houſes, Croſs-Ways, &c. according to
their true Situation, taken at Two Stations, as
you went on the Road; ſo will you (if Care be
taken) have the true horizontal Diſtance of all
thoſe Places within Sight of the Roads, from the
Road itſelf, or from one another.

Secondly, lay down the chief River that runs
through the County, ſo will you have the
Situation of ſeveral more Towns, and other Remarks, as obſerved in your Survey of that River;
and when the main Rivers are done, all the
Branches muſt be protracted with the like Exactneſs; for the main Rivers and Branches being
exactly done, will be a great Ornament to a
County Map.

Thirdly, If the County borders upon the Sea,
firſt protract the Sea-Coaſt exactly, and then take
a Survey of and plott all Rocks, Sands, or other
Obſtacles that lie at the Entrance of any River,
Harbour, Bay or Road upon the Coaſt of that
County, by going out in a Boat to ſuch Sands or
Rocks that make the Entrance difficult; and at
every conſiderable Bend of the Sands, take with
a Sea-Compaſs, the Bearing thereof, to Two
known Marks upon the Shore: And having ſo
gone round all the Sands and Rocks, you may,
upon the Plott before taken of the Coaſt, draw
Lines,

Lines, which fhall interfect each other at every confiderable Point of the Sands; whereby you may give good Directions either for the laying of Buoys, or making Marks upon the Shore, for the Direction of Shipping; and the beft Time to do this, is at low Water, in Spring Tides.

Fourthly, Having truly protracted the principal Roads, Rivers, &c. with the feveral Remarks obferved from thence, you'll find moft of the remarkable Places in the County laid down: But in Order to compleat the Work, look upon fome old Map of the County, and contrive 3 or 4 Market-Towns, or other Towns, to meafure through, that you have not yet laid down, and from thence to other Towns or Villages; and fo do, till you have meafured moft or all of the Roads that lead from Market-Town to Market-Town, taking all the Remarks you can, as you go along; and if you find anything remarkable in the old Map, that you have not taken Notice of, you may go and furvey it. And thus, by Degrees, you may fo finifh a County, that you need not fo much as leave out one Gentleman's Houfe; for fcarce will any thing remarkable efcape coming into your View, either from the Roads, Rivers, or Sea-Coaft.

Fifthly, When you are in a Town, you may place your Inftrument, if you can, upon the Steeple, and from thence take the horizontal Angles to others, by having the horizontal Diftance of thofe 2, from which you take your Angles given; but obferve, all Churches are to be laid down according to their horizontal Diftance one from the other: Therefore, if the Road between them be over Hills of confiderable Height, the Hypothenufal Lines on the Road muft be reduced to Horizontal.

Sixt

Sixthly, All Parks and Forefts muft be truly laid down in the Map, as to their true Bounds and Situation; and all remarkable Lakes and Water: You are likewife to defcribe the Quality of the County, whether it be hilly or woody, placing Hills and Woods in their true Places.

Laftly, Take the true Latitude of the Place, in Three or Four Places of the County; which put down on the Edge of your Map accordingly.

SECT. IV.

General Directions for taking the Ground-plott of a City or other Town.

THE Performance of this Work is very laborious, and you muft be careful to keep the Field-book in a plain and regular Manner, otherwife the Multitude of Obfervations and Offsets will be apt to breed Confufion; but if Care be taken therein you'l find the Work not very hard to be done: One that underftands the fourth Chapter will make no Difficulty of this Section, for the feveral Streets, Lanes, &c. in a City are furvey'd and protracted in the fame Manner as the Lane *Fig.* 22. The feveral Offsets to the Houfes, Churches, &c. all along the Sides of the Street being taken from the main Stationline, running through the middle thereof, in fuch Sort as the Offsets are taken from the Stationline to the Hedges, Gates, &c. on the Sides of the Lane.

The Inftruments for this Purpofe are, 1. the Theodolite as before defcribed, to meafure the feveral Angles made by your Station-lines, as they incline out of one Street into another, and in this Cafe work with the Limb only, but never truft to the Needle, for (befides the Danger of its

being

being attracted) you will find it neceſſary to lay down every Line by ſome other, given in Poſition in the Plott it ſelf, rather than by the Bearings from the Meridian.

2. The Chain; and becauſe the Ground-plott of the Houſes, Pavements, &c. are generally laid out by Foot Meaſure; therefore let every Link thereof be a Foot long, and Fifty of theſe Links will make the Chain of a ſufficient Length, diſtinguiſhed at every Ten Links, by Marks, as *Gunter*'s Chain is: But if the Content of any Part of the Plott be deſired in Acres, you may reduce the Feet in any Line, to Links; and for this Purpoſe the Table in Chap. I. will be a ready Aſſiſtant.

3. The Offset Staff, divided alſo into Feet, 5 of which may make a convenient Length, becauſe you will have Occaſion to meaſure many Paſſages, Alleys, &c. that are not wider; alſo at one End of the Staff, you may have a Piece of about 3 Foot joined, like the Squares of a Drawing-Board; and this will be a Direction to meaſure the Offsets from the Chain, at Right-Angles.

4. The Scale, (or rather a Pair of Beam-Compaſſes) according to the Bigneſs of the Plott, the Protractor, the Drawing-Pen, &c.

Firſt, in one of the principal Sheets, as at ⊙ 1, in the Lane, *Fig.* 22, ſet up a Station-Staff, and ſend another forwards in the Street as far as you can ſee. Then lay the Chain on the Ground exactly in the ſame Direction with the Two Stations, and with the Offset-Staff both to the right and left at Right-Angles from the Chain; meaſure the Offsets as in the Lane; taking Notice at how many Links from the laſt Station each is laid off; and when any of thoſe Offsets reach any

remarkable

remarkable Houfe, &c. or the Corner of a Street, Alley, or Court, enter fuch a Remark againft the refpective Offset, in one of the Outfide Columns of the Field-Book: And in this Manner proceed to the fecond Station.

Set up the Theodolite at the fecond Station, and bring the Index to 360, on the Limb, turning the whole Inftrument about till you fee thro' the Tellefcope the Staff at ⊙ 1; there fix the Inftrument, and then turn about the Index, directing the Tellefcope to another Staff fent forwards in the Street, to the further End thereof, if you can fee fo far; and note in the Field-Book the Angle which the Index cuts on the Limb, with the utmoft Exactnefs: Then proceed with the Chain towards the next Station, as before.

Having in this Manner gone through feveral of the principal, high Streets, that lead through one Part of the Town, it will be convenient, as you pafs along, as often as you come againft any crofs Street, to take a Sight down it; and note the Place or Mark to which the Tellefcope is directed; and alfo at how many Links Diftance from the laft Station the Inftrument is planted, when you thus look into a crofs Street; and note both thefe Places in your Field-Book, or Eye-Draught, with this Mark ⊙; fo that you may be fure to find the Place exactly, when you begin to take your crofs Streets.

It will be convenient, not only to enter your Obfervations in the Field-Book, but alfo to form a Sketch or Eye-Draught of the Work, as you go along, making Lines to imitate the fame; and draw the crofs Streets, Alleys, &c. thereon, in fuch Manner (as near as you can guefs) as you fee thofe crofs Streets to bear from the Place of your Standing in the high Streets; and write the

Name

Name of each Street between the Lines representing the same; and this will be useful when
you come to protract.

Note, Before you begin your Work, it will be
necessary to walk about the Town, and chuse 4 or
5 principal Streets that lead out of one into
another, enclosing between them several By-
Lanes, Alleys, &c. And contrive your first Station
in such a Manner, that when you come round
these 4 or 5 Streets, the last Station-Line may
close exactly on the first Station-Point; and observe, that the fewer Angles you make in going
round these Streets before you close, the better.

This is no more than Surveying a Field; the
main Difficulty will be to find your Stations, when
you come to survey the cross Streets, By-Lanes,
Thoroughfares, &c. between the eminent Streets
that you first went round; but you may help your
self herein, if you lay one End of the Chain at
some Door, or other remarkable Place on the
Right Side of the Street, and draw it in a strait
Line through the Station-Point, to some other
Remark on the other Side of the Street, taking
Notice at how many Links from the Right, the
Chain cut the Station: You may also much help
yourself herein by your Eye-Draught.

When the Station-Line leads you into a Square, you may plant the Theodolite in the Middle
thereof, and from that one Station direct the
Tellescope to the Corners, (very often there are
but four, and the Sides all strait) and measure the
Distances from that Station to the Corners, as in
Chap I. Sect. 3. But if you would take Notice of
particular Houses therein, or if the Sides are very
irregular, then go round it; but Lanes and Alleys
are laid down by Offsets only, from the Station-
Line through the Middle.

Having

Having thus finifhed one Part of the Town or City, you may proceed to another, till the whole be finifhed; but this is a Work that will take up a great deal of Time.

The Manner of Protracting this Work, is the fame as in the preceding Chapters, therefore particular Directions are needlefs; but 'tis beft to protract fo much as you furvey in one Day, before you proceed with more; and for this Purpofe, a Skin of fine, foft Parchment is better than Paper, unlefs the Paper be very fine, and pafted on Cloth or Canvas. The Ground-Plotts of Churches, muft be very exactly taken, and laid down in the fame Manner on the Draught, and fhadowed very deep; the fame of Houfes.

Alfo if you ufe a Protractor that will lay down Minutes, as defcribed in Chap. II. your Work will be more likely to clofe; for you cannot be too curious in obferving and laying down the Angles, efpecially thofe in the principal Streets.

F I N I S.

A P P E N-

APPENDIX

TO THE

Practical SURVEYOR.

CHAP. I.

Of LEVELLING.

SECT. I.

HE Inſtrument moſt approved for this Purpoſe, (a Figure whereof you have in the Beginning of this Book) conſiſts, 1. Of a Braſs Telleſcope of a convenient Length (the longer the Exaɛter, provided the Parts of the Inſtrument that ſupport it, be made proportionably ſtrong:) Within this Telleſcope is fixed an horizontal Hair, and a ſmall Micrometer, whereby Diſtances may be determined at one Station near enough for the Buſineſs of Levelling: Upon this Telleſcope is fixed, with two ſmall Screws, the Spirit Tube, and Bubble therein, which Bubble reſt exaɛtly in the Middle of the Tube, when the Telleſcope is ſet truly level.

2. Under the Telleſcope, is a double Spring, with 2 Screws, by which the Bubble is brought exaɛtly to a Mark in the Middle of the Tube; to which Spring is fixed a Conical Ferril, which is a Direɛtion for the Telleſcope to move horizontally at Pleaſure. There is alſo a three-legged Staff, a Ball Socket, and 4 Screws, to adjuſt the

horizontal

horizontal Motion the fame with that belonging to the Theodolite, before defcribed.

Provide 2 Station-Staves, each 10 Foot long, that may flide one by the Side of the other to 5 Foot, for eafier Carriage; let them be divided into 1000 equal Parts, and numbered at every 10th Divifion, 10, 20, 30, 40, &c. to 100, and from 100, 110, 120, &c. to 200, and fo on till you come to 1000; but every Centeffimal Divifion as 100, 200, 300, to 1000, ought to be expreffed in large Figures, that the Divifions may be more eafily counted; and you may have another Piece 5 Foot long, divided alfo into 500 Parts, to be added to the former, when there fhall be Occafion.

Upon thefe Staves are Two Vanes; made to flide up and down, which will alfo ftand againft any Divifion on the Staff, by the Help of Springs. Thefe Vanes are beft, made 30 Parts wide, and 90 Parts long; let the Faces of them be divided into 3 equal Spaces, by 2 Lines drawn lengthways; let the 2 extreme Spaces be painted white, and let the middle Space be divided alfo into 3 fmaller equal Spaces, and let that in the Middle be painted white, the other 2 black, which will render them fit for all Diftances.

Being thus provided with a good Inftrument, Two Station-Staves, a Chain, and Two Affiftants, you may proceed to your Work; but firft it will be neceffary to make a Trial whether or no your Inftrument be well adjufted.

SECT. II.

How to adjuft the Inftrument.

CHufe fome Field or Meadow, that is nearly level, and fet down the Inftrument about the Middle thereof, and make an Hole in the Ground,

Ground, under the Center of the Inftrument; from which, meafure out in a Right-Line, fome convenient Length, as 20 Chains, and there leave one of your Affiftants with his Station-Staff; then return to the Inftrument, and meafure out the fame Number of Chains, *viz.* 20, the other Way, by the Direction of the Inftrument, and laft Station-Staff, as near a Right-Line as you can guefs, and there leave your other Affiftant with his Station-Staff; fo will the Inftrument and Two Station-Staves be in the fame Line.

Then return to the Inftrument, and fet it horizontal, which is prefently done by the Ball and Socket, and turn the Tellefcope about on its horizontal Motion, to your firft Affiftant, and move the Tellefcope by the Two Screws in the double Spring, till the Bubble refts exactly in the Middle of the Spirit Tube; then obferve where the Hair in the Tellefcope cuts the Staff, and direct your Affiftant to move the Vane up or down, till the Hair cuts the Middle thereof, fo that you may fee as much of the Vane above the Hair as below it, and there give him a Sign to fix it; then direct the Tellefcope towards your fecond Affiftant, and proceed in the fame Manner; fo are the Vanes on each Staff equidiftant from the Center of the Earth.

Remove the Inftrument to that Affiftant which is neareft the Sun, if it fhines, that you may have the Advantage of its Rays upon the other Affiftant's Vane, and there fet down the Inftrument as near the Staff as you can; then having fet the Inftrument horizontal, fo that the Bubble refts in the Middle of the Tube; obferve what Divifion on the Staff is then cut by the Hair in the Tellefcope, above or below the Middle of the Vane, for fo many Divifions muft the other

Affiftant's

Affiftant's Vane be raifed or depreffed, which
direct him to do accordingly.

But becaufe the Inftrument is 40 Chains diftant from the Station-Staff, you muft make an
Allowance for the Earth's Curvature, which by
the following Table you will find to be 16 $\frac{6}{10}$
Parts, therefore let the Vane on the Staff be
raifed 16 $\frac{6}{10}$ Parts.

A T A B L E *of the* Earth's Curvature, *calculated to the Thoufandth Part of a Foot, at the
End of every Chain, from* 1 *Chain to* 40.

Chains	Dec.Foot	Chains	Dec.Foot	Chains	Dec.Foot	Chains	Dec.Foot
1	000	11	013	21	046	31	099
2	000	12	015	22	050	32	106
3	001	13	017	23	055	33	113
4	002	14	020	24	060	34	120
5	003	15	023	25	065	35	127
6	004	16	026	26	070	36	134
7	005	17	030	27	075	37	141
8	007	18	033	28	081	38	149
9	008	19	037	29	087	39	157
10	010	20	041	30	093	40	166

Now direct the Tellefcope to the Vane thus
raifed, and if the Hair cuts the Middle thereof,
while the Bubble refts in the Middle of the
Tube, the Inftrument is right; but if not, you
muft raife or deprefs the Tellefcope by the
Screws in the double Spring, till the Hair cuts
the Middle of the Vane, and then by the Help of
the Screws that fix'd the Tube to the Tellefcope,
move the Bubble till it refts in the Middle of the
Tube: So is the Level adjufted.

S E C T.

SECT. III.

Rules to be observed in Levelling, in Order to find the different Height of any two Places; being useful for conveying Water, cutting Sluices, making Soughs, &c.

S Uppose it was required to know whether Water may be conveyed in Pipes or Trenches, from a Spring Head to any determined Place.

1. At the Head of the Spring set up one of your Station-Staves as nearly perpendicular as you can, and leave with one (whom you may call your first Assistant) proper Directions for Raising or Depressing the Vane on his Staff, according to certain Signs which you (standing at your Instrument) shall give him: Also let him be provided with Pen, Ink and Paper, to note down very carefully the Division on the Staff which the Vane shall cut, when you make a Sign that it stands in its right Position.

2. Carry your Instrument towards the determined Place you are going to, as far as you can see, so that through the Tellescope you may but see any Part of the Staff left behind, when the Instrument is set horizontal; and from that Place send your second Assistant forwards with his Station-Staff with the same Instructions as you gave your first Assistant.

3. Set the Instrument horizontal, by the Help of the Ball, and Socket and 4 Screws; and direct the Tellescope to your first Assistant's Staff, and then by the Help of the Spring-screws bring the Bubble exactly to the Middle of the Tube, and when it rests there, give a Sign for your Assistant to note the Parts of the Staff where the Vane rests.

4. Turn about the Tellefcope to your fecond Affiftant's Staff, and by the Spring-fcrews, as before, fet the Bubble exact: Then direct your fecond Affiftant to move the Vane higher or lower till you fee the Hair in the Tellefcope cut the Middle of the Vane, (but in long Diftances the Hair will almoft cover the Vane; however, let it be fet in fuch Manner that as much may be above the Hair as below it, as near as you can guefs,) and then give him a Sign to note the Divifion on the Staff; and always let your Affiftants note the Divifion cut by the upper Edge of the Vane.

5. Let you firft Affiftant bring his Station-Staff from the Spring Head, and give it to the fecond Affiftant, and let your fecond Affiftant carry it forwards towards the determined Place you are going to, and at a convenient Place erect it perpendicular, whilft your firft Affiftant tarries at the Staff where your fecond Affiftant ftood before.

6. Place your Inftrument between your Two Affiftants, fomewhere about the Middle if you can, and firft direct the Tellefcope to your firft Affiftant's Staff, and when the Tellefcope is levelled to one of the Divifions on the Staff, let him note that Divifion in an orderly Manner under the firft Obfervation; and let your fecond Affiftant do the fame: And in this Manner proceed over Hill and Dale, as ftrait forwards as the Way will permit, to the appointed Place, (by only repeating thefe Directions, tho' it be 20 Miles diftant from the Spring Head;) but in your whole Paffage let this conftant Law be obferved, otherwife great Errors will enfue, (*viz.*) That your firft Affiftant muft at every Station ftand between the Spring-head and your Inftrument,

and

and your second Assistant must always stand between the Instrument and the appointed Place to which the Water is to be convey'd.

Being come to the appointed Place, let both your Assistants give in their Notes, which ought to stand in the Manner and Form following.

First Assistant's Notes.		*Second Assistant's Notes*	
Stations.	Parts.	Stations.	Parts.
⊙ 1	1019	⊙ 1	330
⊙ 2	512	⊙ 2	540
⊙ 3	737	⊙ 3	1337
⊙ 4	40	⊙ 4	742
⊙ 5	1495	⊙ 5	30
⊙ 6	1475	⊙ 6	32
⊙ 7	1430	⊙ 7	30
⊙ 8	1149	⊙ 8	227
Sum.	7857	*Sum.*	3268

These Notes were collected from Observations made with such an Instrument, as before described, at several Stations between the Ground at the North-gate of *Hanover-Square*, and the Surface of the square Pond by the *New-River Head*, near *Islington*. The first Assistant's Notes, when added together, amount to 7857; the second Assistant's 3268, the Difference 45, 89 Parts; that is almost 46 Foot; and so much is the Pond higher than the Ground of that Part of the Square where the first Station-staff was planted.

The following Observations were repeated in the Afternoon of the same Day, at quite different Stations; from the Pond before-mention'd, to the said North-gate of *Hanover-Square*; and then the two Assistant's Notes stood in the following Manner.

First Assistant's Notes.		Second Assistant's Notes	
Stations.	Parts.	Stations.	Parts.
☉ 1	290	☉ 1	1278
☉ 2	36	☉ 2	1515
☉ 3	77	☉ 3	1395
☉ 4	68	☉ 4	1500
☉ 5	58	☉ 5	74
☉ 6	1243	☉ 6	38
☉ 7	998	☉ 7	468
☉ 8	437	☉ 8	774
☉ 9	306	☉ 9	1066
Sum.	3513	*Sum.*	8108

These Notes as observed in the Afternoon, being added together, and the lesser subtracted out of the greater, the Difference is 45, 95 Parts, which very nearly agrees with the former Observations; being but $\frac{6}{100}$ of a Foot Difference, which is inconsiderable.

Note, If from the first Assistant's Staff you measure any Number of Chains towards the Place you are going to, suppose 10, and there set down the Instrument, and then measure 10 Chains forwarder and there place the other Station-Staff, you will have no Occasion to make any Allowance for the Curvature of the Earth, because the Instrument being planted in the Middle of the Distance between the Station-staves, the Errors mutually destroy each other.

But this measuring of the Distances with the Chain, or otherwise, is very tedious, and indeed impracticable in many Cases, unless you make a Multitude of Stations: So if the Way between

two

two determined Places, whose different Height you would know, lies over Hills and Dales, as *Fig.* 32, you must in that Case make four or five Stations (otherwise you will not be able to see any Part of the Staff, when the Inftrument is set horizontal,) which might as well be done at one, (as in the foregoing Obfervations,) in the following Manner.

SECT. IV.

How to make Allowance for the Curvature of the Earth, when the Station-ftaves are planted at unequal Diftances from the Inftrument.

SUppofe the Inftrument was planted on the Eminence between two Valleys *a*, and *b*, and the firft Affiftant with his Station Staff, ftanding at *c*, and the fecond at *d*, and it is required to know the different Height of the Hills *c*, and *d*.

Firft fet the Inftrument horizontal; and then direct the Tellefcope to the firft Affiftant's Staff at *c*, and by the Spring-fcrews fet the Bubble exact, obferving where the Hair cuts the Staff, and by Signs caufe him to move the Vane higher or lower till the Hair cuts the Middle thereof; and then give him a Sign to note the Divifion cut by the upper Edge of the Vane, which fuppofe 104 Parts from the Ground, and by the Micrometer in the Tellefcope, I find the Diftance from the Inftrument to the Staff at *c*, to be about 10 Chains.

Then I direct the Tellefcope to *d*, and procede in the fame Manner as before, and find that the Hair cuts 849 Parts from the Ground, and by the Micrometer the Diftance *d*, is determined to be about 35 Chains.

Next I look into the Table of Curvature and find

find againſt 10 Chains, 1 Part to be deducted for the Curvature of the Earth at that Diſtance; ſo will the firſt Aſſiſtant's Note be made 103 Parts.

Alſo againſt 35 Chains I find 12 $\frac{7}{10}$, which deducted out of 849, there remains 836 $\frac{3}{10}$ Parts which muſt be noted by the ſecond Aſſiſtant.

Now if 103, as noted by the firſt Aſſiſtant, be ſubtracted from 836 $\frac{3}{10}$, as noted by the ſecond, the Remainder will be 733 $\frac{3}{10}$; and ſo much the Hill *c*, is higher than the Hill *d*: But if you have not the Table of Curvature at Hand, then you may find the Allowance that is to be made at any Diſtance, by this Rule.

Multiply the Square of the Diſtance in Chains by 31, and divide the Product by 300000.

In this Manner making Allowance for the Curvature of the Earth, you may ſend a Station-ſtaff forwards half a Mile, or farther from the Inſtrument; and take a Sight over ſeveral Valleys at once, the horizontal Diſtance in this Caſe being only regardable.

Note, When Water is to be brought to any appointed Place; there muſt be an Allowance of 4 $\frac{1}{2}$ Inches for every Mile, more than the ſtrait Level, for the Current of the Water; but if the Spring-head be much higher than the appointed Place, ſo that the Water will have too violent a Current, the Pipes may be laid one up and another down; and inſtead of being laid in a ſtrait Line, the Water be brought in a crooked or winding Way.

C H A P.

CHAP. II.

Shewing the Use of the Theodolite, *in Drawing Buildings,* &c. *in* Perspective.

SECT. I.

WHEN a Building is to be drawn upon a Perspective Plane (or Picture,) the Representation of the several Objects ought to be delineated thereon according to their Dimensions and different Situations, in such Manner that the said Representations may produce the same Effects on our Eyes as the Objects whereof they are the Pictures.

But without Mathematical Rules this Representation cannot well be found; for when Objects are drawn by only Viewing or Looking at them, their true Representations will often be miss'd; whereas by the following Method they may always be obtained.

For all Objects appear such as the visual Angle under which they are seen; which Angle is taken at the Eye, where the Lines meet that do comprise the Object; that is to say, an Object seen in a great Angle, will appear great; and another seen in a little Angle, will appear little; which is the principal Thing to be observed in Perspective.

So

So the windows 6, 7, 8, *Fig.* 33, muſt be drawn on the Perſpective Plane of different Dimenſions (altho' on the Building one of 'em is really as big as the other) according to the Angle which the Rays from their Extremities make with the Eye at *z*.

Objects of equal Bigneſs appear greater or leſs, according to their Diſtance from the Beholder's Eye; ſo the Windows 6 and 8, are really one as big as another on the Geometrical Plane; but the Window 6, at the End of the Building being nearer the Eye at *z*, than the Window 8 on the Front, it muſt be made ſo much larger on the Perſpective Plane, as the Window 6 is to that marked with 8.

Therefore, if the Angles, under which Objects appear, be given, thoſe Objects may be drawn on the Perſpective Plane (or Paper) according to their Dimenſions and different Situations, in the ſame Form as they appear to the Beholder at any Diſtance.

The Figure on the Geometrical Plane (or Building) are compos'd either of ſtrait Lines or Curves: Now to find the Repreſentation of a ſtrait Line, its Extremities need only be ſought: And to find the Appearance of a Curve, we need only to find the Place of ſeveral Points therein. And hence it follows, that the whole Buſineſs of Perſpective conſiſts in finding only the Place of a Point.

But theſe Points can't be determined, unleſs by the Interſection of Right Lines. And the Reaſon of theſe Sections is, That one Line can determine nothing: Therefore it is neceſſary, that there be Two of them, which divide themſelves, (forming an Angle) for to have the Place of a Point, as will be ſeen in the following Example.

For

For having noted the Obſervations made by the Theodolite, the Plan of any Building may be drawn in Perſpective, without meaſuring ſo much as one Line; or coming nearer the Building than where the Inſtrument is planted.

SECT. II.

Let Fig. 33. repreſent the Building as viewed from Z, being the Place from which the Proſpect is deſired to be taken.

THE Inſtrument being planted at *z*, and the Staves made to ſtand firm on the Ground, I ſet the Inſtrument exactly level; and with the Index at 360, and the Quadrant at o degrees, direct the Telleſcope to ſome Part of the Building, as to *o*, by turning about the whole Inſtrument, and there ſcrew it faſt, that it ſtir not out of this Poſition, till the ſeveral Obſervations be finiſhed.

The Inſtrument being ſet level, the Index, when turned round on the Limb, carries the Telleſcope in a Line Parallel to the Horizon, as *x, y*: And the Quadrant elevated or depreſſed, moves always in a Circle vertical thereto, as *w, z*.

Now take the Pin out of the Quadrant, and with one Hand move the Index on the Limb, and with the other elevate or depreſs the Telleſcope as there ſhall be Occaſion, till you ſee the croſs Hairs therein cut any Point on the Building; and then note down on a Piece of Paper, the Degrees and Minutes which the Index cuts on the Limb in one Column, and call thoſe the horizontal Angles: Likewiſe note the Degrees and Minutes cut by the Quadrant in another Column, and call thoſe the Vertical Angles.

So the Telleſcope being directed to the Point

H

a, the Index then cuts 7°, 25′, and the Quadrant 19°, 30′; and thofe Obfervations when protract-ed, will give the Point *a*.

Likewife I make Obfervation of the Point *b*; and then deprefs the Tellefcope to the Bottom of the Building at *c*, and the Index then cuts the fame Angle on the Limb as at *b*, and the Qua-drant 8°, 30′: But this Angle of Depreffion muft be marked with ∧, or fome fuch Mark to diftin-guifh it from the Angles of Elevation, that in Protracting that Point, it may be known to be under the Horizon, or the Line *x*, *y*.

When the Inftrument is planted at a confide-rable Diftance from the Building, the Ground there may be higher or lower than any Part of the Building: And then all the Points will be above or under the Horizon; and in fuch Cafe there will be no Occafion for this Diftinction.

In the fame Manner I make Obfervation of fo many Points on the Right Side of the Houfe as is convenient; but when the Tellefcope is directed to the Point *m*, on the Left Side, the Index cuts 340°, 40′.

Now this number 340°, 40′, muft not be noted for the horizontal Angle, but its Complement to 360 (*viz.*) 19°, 20′, by fubtracting 340°, 40′, out of 360; but if the Degrees be numbred by fmall Figures from 360, the contrary Way, as 10, 20, 30, *&c.* to 60, or further, as may conveniently be done, the Numbers will encreafe from 360, both to the Right and Left; and then the Index will always cut the Number denoting the horizontal Angle, in the fame Manner as the Quadrant.

Having obferved the Point *m*, the Index re-maining at the fame Angle on the Limb, I deprefs the Tellefcope to the Points 4, 3, 2, 1, and note the Degrees *&c.* cut by the Quadrant;

which

which when protracted, will give the Breadth of the Facies and the Distance one from another.

Next I observe the Points of the Window *e*, *i*, *u*, in the Left Wing of the Building; and because these Remarks are on the Left Side of the Building, therefore I note them by such Names as I call the several Points I look at, (instead of the Letter, *a*, *b*, *&c.*) on the Left Side of the Column of Observations, (*viz.*) contrary to that Part of the Limb where the Index cuts, (which remember;) for when the Index is turned from 360, on the Limb towards the Right Hand, the Tellescope moves towards the Left: And these Remarks thus noted, must be protracted on the Left Side of the Vertical Line *w*, *z*, *Fig.* 33.

In making these Observations, 1. Set the Instrument level in that Place from which the Prospect is desired to be made; and with the Index at 360, direct the Tellescope to some remarkable Place about the Middle of the Building, and there fix the Instrument.

2. The Remarks on the Right Side of the Building, enter in the Column of Observations on the right Hand; *& è contra.*

3. If there be Angles both of Elevation and Depression, mark the Angles of Depression with $\wedge$.

The Observations of most of the Points, that need to be taken of *Fig.* 33, in order to protract or draw the same in Perspective, are inserted in the following Table: And observe, that if the Building be regular, there will need but few Points to be given; for, where you have the Height and Breadth of one Window given, with its Distance from the next, the whole Row may thereby be drawn, being all of the same Dimensions, but Objects more irregular must be

drawn

drawn by obferving fo many Points therein, as fhall be neceffary: But Practice in this Cafe is the beft Guide.

SECT. III.

The Manner of Protracting thefe Obfervations in Order to find the Points of the Building, Fig. 33.

Horizon. Angles	Vertical Angles	
7° : 25′	19° : 30′	a
11 : 30	16 : 30	b
11 : 30	8 : 30	c $\wedge$
19 : 20	26 : 50	d
19 : 20	13 : 30	e $\wedge$
38 : 00	26 : 50	f
38 : 00	13 : 30	g $\wedge$
4 19 : 20	26 : 50	
	25 : 30	
3 ———	18 : 40	
2 ———	9 : 35	
$\wedge$: 1 ———	1 : 00	
i 18 : 30	19 : 00	
e 18 : 30	22 : 40	
u 16 : 50	17 : 50	

FIRST draw a right Line *x, y, Fig.* 33. for the Horizontal-line; and at right Angles therewith draw another Line *w, z,* which reprefents the Vertical-line.

Set off the Points of Diftance from *o, (viz.)* from that Point where *x, y,* and *w, z,* interfect one another: And according to what Bignefs you would

would have the Plan of the Building be, make the Diftance bigger or lefs. If you would have the Draught large, make the Diftance large: *Et è contra.* Therefore fet one Foot of the Compafs at *o*, and with the fame Extent mark the Points of Diftance *x, y, z.*

The Horizontal-angles muft be drawn from the Point *z*, to the horizontal Line *x, y*; and the vertical Angles from the Points *x*, or *y*, (according as the Remarks are noted on the right or left Side of the Columns) to the Vertical-line *w, z.*

The Index being at 360, and the Quadrant at *o*, when the Crofs-hairs in the Tellefcope cut the Point *o*, on the Building: Therefore the Point *o*, fhall be the firft Point of Sight on the Profpective-plan.

By the Table of Obfervations, I find that the Index cuts 7°. 25′. on the Limb; and the Quadrant 19° 30′, when the Tellefcope was directed to the Point *a*: Therefore lay the Center of the Protractor to *z*: And becaufe the Letter *a* is noted on the right Side of the Columns, lay the Limb on the Right Side of the Line *w, z*, the Diameter coincident therewith; and againft 7°, 25′, make a Mark clofe by the Limb of the Protractor.

Lay the Edge of a ftrait Ruler to the Point of Diftance *z*, and to that Point 7°, 25′; and where the Edge cuts the Horizontal Line, make a Mark.

Lay the Center of the Protractor to the Point of Diftance *y*, (becaufe *a* is noted on the Right Side of the Columns) the Diameter coincident with the Line *x, y*; and againft 19°, 30′, on the Limb, make a Mark.

Lay a ftrait Ruler to that Mark, and the Point of Diftance *y*; and where the Edge cuts the Vertical Line *w, z*, make a Mark at *r*.

Laftly, Lay a Parallel Ruler to the Horizontal
Line

Line *x*, *y*, and move it Parallel thereto, till the Edge cuts the Point *r*, in the Vertical Line; and with the Compaſs Point draw the obſcure Line *r* 5.

Then lay the Parallel Ruler to the Vertical-Line *w*, *z*, and move it Parallel thereto till the Edge cuts the Point *t*, in the horizontal Line *x*, *y*, and by the ſame Edge draw an obſcure Line *t* 9; ſo ſhall the Interſection of theſe Two Parallels determine the Place of the Point *a*, which was ſought.

In the ſame Manner may the Point *b*, or any other Point be found: and then theſe Points joined with Right Lines, ſhall repreſent the Lines on the Building, and bear an exact Proportion thereto, according to the Rules of Perſpective.

The Point *c* is found in the ſame Manner as *a*, only becauſe the horizontal Angle is the ſame with the Point *b*, you have no more to do but continue a ſtrait Line from *b*, through the Point *x*, in the Horizontal Line, Parallel to *w*, *z*; and then lay the Center of the Protractor to *y*, with the Limb downwards, becauſe *c* is marked with /\ (*i. e.*) under the Horizon; and draw the Vertical Angle 8°, 30′, to 8 in the Vertical Line; ſo ſhall a Line drawn Parallel to *x*, *y*, from the Point 8, cut the obſcure Line *b*, *c*, at *c*, the Point ſought.

The Points *h*, *k*, *l*, *m*, *n*, *p*, *q*, on the Left Side of the Building, *Fig.* 33, have the ſame Angles with *a*, *b*, *c*, *d*, *e*, *f*, *g*, on the Right, and therefore protracted in the ſame Manner; except this Difference, that becauſe the Points *h*, *k*, *l*, &c. are on the Left Side of the Building, therefore the ſame Points muſt be found on the Left Side of the Vertical Line *w*, *z*, and the Protractor laid to the Point of Diſtance *x*; but the Horizontal Angles are all laid off from the ſame Point of Diſtance *z*.

Obſerve,

Obferve, That in Protracting thefe Points, 'tis convenient, that the Numbers of the Semicircular Protractor fhould be made to encreafe from the Diameter both Ways, that the Numbers may be counted thereon, both to the Right and Left: And then in Protracting any Point on a Building, 1. Draw the Horizontal Angle from the Point of Diftance *z*, to the Horizontal Line *x*, *y*, as to *t*. 2. Draw the Vertical Angle to the Vertical Line *w*, *z*, as to *r*. 3. Draw Lines Parallel to *w*, *z*, and *x*, *y*, through the Points *t* and *r*; fo fhall the Interfection of the Two Parallels give the Point fought.

But thefe Points are found with much greater Expedition, if the Paper on which you draw the Plan of the Building, be faftened to a Drawing-Board, and the Angles laid down the Sector in the following Manner:

For Example: Suppofe the Point *a*, *Fig.* 33, was fought.

Firft, Draw Two Lines by the Side of the Tee, croffing one another at Right Angles, as *x*, *y*, and *w*, *z*, *Fig.* 33.

Take between the Points of the Compaffes the Diftance *z*, *o*, and let the Sector be opened to the fame Extent, by fetting one Foot of the Compaffes at the End of the Tangent-Line, at 45, on one Side of the Sector, and let the other fall at the other End of the Tangent-Line at 45, of the other Side of the Sector.

The Sector remaining at this Extent, fet one Foot of the Compaffes in the Tangent-Line on one Side of the Sector at 7°, 25′, the Horizontal Angle, as in the Table; and let the other fall at 7°, 25′, on the other Side; this Diftance fet from the Point of Sight *o*, in the Horizontal-Line *x*, *y*, to *t*.

In

In the fame Manner take off from the Sector the vertical Angle 19° 30' which fet on the vertical Line *w, z*, from *o* to *r*.

Laftly, lay the Tee on the Drawing-board, parallel to *w, z*, fo that the Edge cut through the Point *t*, and draw the Obfcure-Line *t*, 9.

Lay the Tee to the other Side of the Drawing-Board parallel to *x, y*: And the Edge cutting through the Point *r*. Draw the Obfcure-Line *r*, 5, fo fhall the Interfection of thefe two Lines *t*, 9, and *r*, 5, give the Point *a* which was fought.

In the fame Manner may any other Point be found in as little Time as it could be obferved by the Theodolite; but if you have not a Drawing-board nor Parallel-ruler, you may put the Paper on the plain Table and by the Edge of the Index laid on the equal Divifions, draw the Parallels; but a Drawing-board is better.

Having found the Points *m* and *k*, both denoting the upper Part of the Facies, if you lay a Ruler to thefe two Points, and continue a ftrait Lines till it cuts the Horizontal-line *x, y*, as at *f*, that fhall be the accidental Point, (or, as the Draughts-Men fometimes call it, the vanifhing Point,) which being found, you may from thence draw right Lines to any other Points on the Draught which were viewed obliquely from *z*; (and therefore the Figures on that Part of the Building muft be made inclined on the Draught,) and thereby find the Abridgment of all the Lines parallel to the Horizon on the Building or geometrical Plan; (which is fuppofed parallel to the perfpective Plan or Picture.)

So when you have protracted the vertical Angles of 4, 3, 2, 1, *Fig.* 33. and thereby found thofe Points, You may lay a Ruler to each of them, and the accidental Point *f*, and thereby draw the

Facies

Facies on the Wings of the Building according to their Breadth and Distance from one another on the perspective Plane or Draught.

In like Manner having found the Points *e*, *i*, and *u* of the first Window, you may from *e* and *i*, draw Lines to the accidental Point *ſ*, which will give the Bottoms and Tops of all that Row: And then you have nothing to do, but find their Breadth and Distance; and by these Directions draw all the Windows on that Wing of the Building.

If a Statue, Coat of Arms, or other Object was placed at *o*, *Fig.* 33. and it was desir'd to place the same (or another,) a good Deal higher, as at *r*; but so, that the Object when placed at *r* should appear full as big, as when at *o*; being viewed from *z*.

Observe with the Theodolite, the Angles under which the Object appears at *o*, as if it was a Statue, observe the Height from the Feet to the Head, *&c.* and note the Angles with proper Remarks on a piece of Paper; and then by directing the Tellescope to *r*, and setting the Quadrant and Index to the same Angles, you may give Directions how to make the Object at *r*, of such Dimensions as being viewed from *z*, will appear of the same Magnitude (or natural Height) with that at *o*; *& vice verſa.*

The same may be done, if Objects are desired to be placed at a Distance, to appear of the same Size as those that are nearer; with several other Problems to be performed by this Instrument, which the Ingenious will find out in the Use thereof: But I have already exceed what I intended on this Head, and shall only add two or three Astronomical Problems, which the Surveyor perhaps may find very useful in Practice.

C H A P.

CHAP. III.

PROBLEM I.

*How to find a True Meridian-Line, by Obferving
with the Theodolite.*

HE beft Time to make the Obfer-
vations, are in a clear Day, about 3
or 4 Hours before and after Noon.

In the Morning, having fet the
Inftrument exactly level, move the
Index Horizontally, and the Quadrant Vertical-
ly, till through the Tellefcope you fee the crofs
Hairs in the Center of the Sun: Then obferve
what Degrees and Minutes are cut by the Index,
fuppofe 3°, 25′, which note in a Piece of Paper, as
alfo the Angle of Elevation cut by the Quadrant.

About fo many Hours after Noon, obferve exactly, that the Quadrant be fet to the fame Angle
of Elevation as in the Morning; and then move
the Index on the Limb till you fee the crofs Hairs
cut the Center of the Sun, as in the Morning;
and note the Degrees and Minutes which the
Index then cuts on the Limb, fuppofe 64°, 37′.

But Note, 'tis convenient in the Morning to
make 3 or 4 Obfervations 5 or 6 Minutes from
one another; becaufe in the Afternoon you muft
wait till the Sun falls into the fame Altitude as it
had when you made the Obfervation in the
Morning, (the Quadrant remaining at the fame
Angle at both;) and if it fhould happen to be
obfcured

obfcured by Clouds at that Inftant, your Labour
will be loft for that Day, having made but one
Obfervation in the Morning.

Now, if from 64°, 37′, the Evening Obfervation on the Limb, you fubtract 3°, 25′, the
Morning Obfervation, the Remainder will be
61°, 12′, the half of which is 30°, 36′; to this half
Sum 30°, 36′, add the Morning Obfervation 3°,
25′, and the Sum will be 34°, 1′.

Laftly, the Inftrument remaining in the fame
Pofition, bring the Index on the Limb to 34°, 1′,
and the Quadrant and Tellefcope will be exactly
in the Plane of the Meridian: But if the Obfervation on the Limb in the Morning, exceed that
in the Afternoon, you muft add to the Afternoon
Obfervation 360, and work in like Manner; and if
the Remainder fhould exceed 360, you muft
fubtract 360 therefrom.

Now obferve what Point (on fome firm Wall of
a Building) is cut by the crofs Hair in the
Tellefcope, there caufe a good Mark to be fixed,
or caufe a Pillar with a Mark thereon to be fet up
in the Direction of the Tellefcope: Alfo take
Notice, If you could place the Mark a Quarter or
Half a Mile diftant from the Inftrument, it is
better than if it was nearer. And in making thefe
Obfervations, you ought to be very exact;
becaufe when a Meridian-Line is once well fixed,
it is very ufeful for divers Purpofes.

Obferve, When the Sun in near the Tropicks,
the Meridian-line may be found well enough by
obferving as aforefaid: But when it is near the
Equinox, there will be fome Variation; becaufe
the Sun's Declination is greater or lefs at different Times in the fame Day: And confequently
when in equal Altitudes, has different Azimuths.
Therefore the Meridian-line may be more truly
found,

found, By the *Pole Star*.

The following Table ſhews the Time from Noon, when the Pole Star makes the greateſt Angle from the Meridian-Line towards the Eaſt, for any Time in the Year.

January.			*February.*			*March.*			*April.*		
D.	H.	M.	D.	H.	M.	D.	H.	M.	D.	H.	M.
1	23	16	5	20	52	5	19	8	2	17	27
8	22	51	12	20	26	12	18	43	9	17	0
15	22	17	19	19	59	19	18	17	16	16	35
22	21	48	26	19	33	26	17	52	23	16	7
29	21	20							30	15	40

May.			*June.*			*July.*			*Auguſt.*		
7	15	13	4	13	19	2	11	23	6	9	6
14	14	45	11	12	49	9	10	55	13	8	40
21	14	16	18	12	20	16	10	26	20	8	14
28	13	48	25	11	52	23	9	59	27	7	50
						30	9	33			

September.			*October.*			*November.*			*December.*		
3	7	23	1	5	42	5	3	26	3	1	26
10	6	58	8	5	16	12	2	57	10	0	54
17	6	32	15	4	50	19	2	27	17	0	24
24	6	8	22	4	22	26	1	57	24	23	53
			29	3	54				31	23	23

And, to find the Time that the Pole Star will make the greateſt Angle on the Weſt of the Meridian; add 11 h. 84 m. to the Time found in the Table: Alſo *Note*, The Star comes to the ſame Place about 4 Minutes ſooner every 24 Hours, than it did the Day before.

The

The following Table shews the greatest Angle which the Pole Star makes with the Meridian in any of these Latitudes, (viz.)

Latitude.		Angles.	
Deg.	Min.	Deg.	Min.
49	00	3	53
49	30	3	55
50	00	3	57
50	30	4	00
51	00	4	03
51	32	4	6
52	00	4	9
52	30	4	12
53	00	4	15
53	30	4	18
54	00	4	21
54	30	4	24
55	00	4	27
55	30	4	30
56	00	4	34

The Time that the Pole Star comes to the East or West of the Meridian, and the greatest Angle which it makes therewith, being found by the preceding Table: Set the Theodolite horizontal, and bring the Index to 360 on the Limb; then turn the whole Instrument about, and elevate or depress the Tellescope, till you see the Pole Star in the Intersection of the Hairs therein, and there screw the Instrument fast: Then (if the Observation was made in the Latitude of *London*, 51°, 32′) move the Index on the Limb 4°, 6′, (as by the Table) towards the Right Hand or Left, according as the Star is Westward or Eastward: And the Tellescope will be set exactly in the Plane of the Meridian.

PROB

PROBLEM 2.

How to find the Latitude of any Place, by the Theodolite.

THE Inftrument being fet level, bring the Quadrant and Tellefcope into the Plane of the Meridian, and let the Index remain at the fame Angle on the Limb, then elevate or deprefs the Tellefcope towards the Sun, at fuch Time as you think it is near the Meridian, until you fee the crofs Hairs in the Center thereof, dividing it as it were into four equal Quarters; and obferve exactly what Degrees and Minutes are then cut on the Quadrant, fuppofe 42°, 15′, which note for the Sun's Meridian Altitude.

By an Ephemeris, you may find the Sun's Declination for the fame Day, fuppofe 3°, 47′, which if it be North Declination, fubtract it from 42°, 15′, the Meridian Altitude, and the Remainder will be 38°, 28′, the Co-latitude.

But if the Sun hath South Declination, add it to the Meridian Altitude, and the Sum will be the Co-latitude; which fubtracted from 90°, gives the Latitude of the Place.

PROBLEM 3.

How to find when the Sun or any of the Stars are upon the Meridian: And the exact Limits of the Natural Day.

HAving the Co-latitude of the Place, by the laft Problem, and the Declination of the Sun given; add the Declination, if North, to the Co-latitude; but if South, fubtract it, and the Remainder will be the Sun's Meridian Altitude for the Day, as aforefaid, which fuppofe to be 42°,

42°, 15′.

Set the Quadrant to 42°, 15′, and the Tellefcope will be elevated to the Meridian Altitude
of the Sun; then note the Inftant of Time by a
Watch or Pendulum-Clock, when through the
Tellefcope (remaining at the fame Angle) you
fee the crofs Hairs cut the Center of the Sun; for
at that Time is the Sun upon the Meridian.

And if you proceed in like Manner the next
Day, you will have the exact Limits of the Natural Day, which muft exceed or want fo many
Seconds of 24 Hours, by your Clock or Watch, as
appears by the Equation-Table for the Day, if
your Clock or Watch goes right.

In the fame Manner, you may obferve when
any Star comes to the Meridian 3 Minutes, 56
Seconds and a half, fooner the fecond Night than
it did the firft, your Pendulum-Clock or Watch
keeps true Time, *& è contra.* Alfo if you fubtract
3 Minutes, 56 Seconds and a half, for each Night
after that on which you made the firft Obfervation, you will have the true Time of that Star's
coming to the Meridian for each Night following.

And thus may a Pendulum-Clock or Watch be
adjufted to the Mean Motion of the Sun.

PROBLEM 4.

How the Azimuth and Altitude of any of the fixed Stars are found by the Theodolite.

THE Inftrument being fet level, and exactly in the Plane of the Meridian, and there fixed, if you direct the Tellefcope to any Star, its Azimuth is fhewn by the Index on the Limb; and the Altitude by the Quadrant both at the fame Time.

F I N I S.

Oblemens and Gentlemens Eftates Sur-
vey'd and Valu'd, in order to their
Improvement, and Books of Maps with the
Particulars, drawn from the faid Surveys;
fhewing the feveral Sorts of Lands, and the
Yearly Value of each Parcel in the Poffeffion
of the feveral Tenants, whether Tenants at
Will, or upon Lives, &c.

By { *Edward Laurence,*
AND
William Gardiner.

Who may be heard of, at Mr. *Jonathan
Siffon*'s, Mathematical Inftrument-maker, in
the *Strand*, or at Mr. *Mead*'s, a Goldfmith,
near *Temple-Bar*.

N. B. The faid *Edward Laurence* having
had long Experience in County Bufinefs,
undertakes to value Eftates already Survey'd,
either for the Buyer or Seller, &c. And draws
out proper Covenants to oblige Tenants to
keep their Farms up to the due Courfe of
Husbandry.

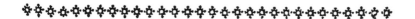

Some BOOKS *printed for* J. HOOKE, *at the* Flower-de-luce, *in* Fleet-ſtreet.

I. THE Compleat Meaſurer; or the whole Art of Meaſuring, in Two Parts. The firſt Part teaching Decimal Arithmetick, with the Extraction of the Square and Cube Roots. And alſo the Multiplication of Feet and Inches, commonly call'd Croſs Multiplication. The Second Part teaching to Meaſure all Sorts of Superficies and Solids, by Decimals, by Croſs-Multiplication, and by Scale and Compaſſes: Alſo the Works of ſeveral Artificers relating to Building; and the Meaſuring of Board and Timber. The Second Edition. By William Hawney, Pr. 2 s. 6 d.

II. Navigation Improv'd: In Two Books. 1. Containing an Exact Deſcription of the Fluid Quadrant for the Lati-tude, *&c.* 2. An Eſſay on the Diſcovery of the Longitude, by a new Invention of an Everlaſting Horometer. Founded on the moſt unerring Principles of Nature. With Copper-Plates of the Inſtruments, *&c.* By Captain *Jacob Row.* Price 2 s.

III. An Exact Survey of the Tide. Explicating its Pro-duction and Propagation, Variety and Anomaly, in all Parts of the World, eſpecially near the Coaſts of *Great-Britain* and *Ireland*: With a Preliminary Treatiſe concerning the Origin of Springs, Generation of Rain, and Production of Wind. By *E. Barlow*, Gent. With curious Maps. Pr. 5 s.

IV. The Dying Speeches and Behaviour of the ſeveral State Priſoners that have been Executed the laſt 300 Years. With their ſeveral Characters, from the beſt Hiſtorians, as *Cambden, Spotſwood, Clarendon, Sprat, Burnet, &c.* By Mr. *Salmon.* In 8*vo.* Pr. 6 s.

V. The Moral Characters of *Theophraſtus.* Tranſlated from the *Greek*; with Notes. To which is prefix'd a Critical Eſſay on Characteriſtick-Writings. By *Hen. Gally*, M.A. Pr. 3 s. 6 d.

TABLES,

SHEWING THE

ALTITUDE and DIAMETER

OF ANY

OBJECT

To the hundredth Part of a Foot,

Anſwering to every Tenth Part of a Degree,
throughout the DOUBLE SEXTANT; obſerved
by the New THEODOLITE, from a Station of
ten, twenty, thirty, &c. Feet Diſtant.

AS ALSO

The Fourth Part of the Girt of any Timber-Tree
ſtanding.

D. T.	Parts.	D. T.	Parts.	D. T.	Parts.	D. T.	Parts.	D. T.	Parts.	D. T.	Parts.
0	0	5:0	87	10:0	176	15:0	268	20:0	364	25:0	466
1	1	1	89	1	178	1	270	1	366	1	468
2	3	2	91	2	180	2	272	2	368	2	470
3	5	3	93	3	182	3	273	3	370	3	473
4	7	4	95	4	183	4	275	4	372	4	475
5	9	5	96	5	185	5	277	5	374	5	477
6	10	6	98	6	187	6	279	6	376	6	479
7	12	7	100	7	189	7	281	7	378	7	481
8	14	8	102	8	191	8	283	8	380	8	483
9	16	9	103	9	192	9	285	9	382	9	485
1:0	17	6:0	105	11:0	194	16:0	287	21:0	384	26:0	488
1	19	1	107	1	196	1	289	1	386	1	490
2	21	2	109	2	198	2	290	2	388	2	492
3	23	3	110	3	200	3	292	3	390	3	494
4	24	4	112	4	202	4	294	4	392	4	496
5	26	5	114	5	203	5	296	5	394	5	498
6	28	6	116	6	205	6	298	6	396	6	501
7	30	7	117	7	207	7	300	7	398	7	503
8	32	8	119	8	209	8	302	8	400	8	505
9	33	9	121	9	211	9	304	9	402	9	507
2:0	35	7:0	123	12:0	212	17:0	306	22:0	404	27:0	509
1	37	1	125	1	214	1	308	1	406	1	512
2	38	2	126	2	216	2	309	2	408	2	514
3	40	3	128	3	218	3	311	3	410	3	516
4	42	4	130	4	220	4	313	4	412	4	518
5	44	5	132	5	222	5	315	5	414	5	521
6	45	6	133	6	223	6	317	6	416	6	523
7	47	7	135	7	225	7	319	7	418	7	525
8	49	8	137	8	227	8	321	8	420	8	527
9	51	9	139	9	229	9	323	9	422	9	529
3:0	52	8:0	140	13:0	231	18:0	325	23:0	424	28:0	532
1	54	1	142	1	233	1	327	1	426	1	534
2	56	2	144	2	234	2	329	2	428	2	536
3	58	3	146	3	236	3	331	3	431	3	538
4	59	4	148	4	238	4	333	4	433	4	541
5	61	5	149	5	240	5	334	5	435	5	543
6	63	6	151	6	242	6	336	6	437	6	545
7	65	7	153	7	244	7	338	7	439	7	547
8	66	8	155	8	246	8	340	8	441	8	550
9	68	9	157	9	247	9	342	9	443	9	552
4:0	70	9:0	158	14:0	249	19:0	344	24:0	445	29:0	554
1	72	1	160	1	251	1	346	1	447	1	556
2	73	2	162	2	253	2	348	2	449	2	559
3	75	3	164	3	255	3	350	3	451	3	561
4	77	4	165	4	257	4	352	4	454	4	563
5	79	5	167	5	259	5	354	5	456	5	566
6	80	6	169	6	260	6	356	6	458	6	568
7	82	7	171	7	262	7	358	7	460	7	570
8	84	8	173	8	264	8	360	8	462	8	573
9	86	9	174	9	266	9	362	9	464	9	575
5:0	87	10:0	176	15:0	268	20:0	364	25:0	466	30:0	577

D. T.	Parts.	D. T.	Parts.	D. T.	Parts.	D. T.	Parts.	D. T.	Parts.	D. T.	Parts.
30:0	577	35:0	700	40:0	839	45:0	1000	50:0	1192	55:0	1428
1	579	1	702	1	842	1	1003	1	1196	1	1433
2	582	2	705	2	845	2	1007	2	1200	2	1439
3	584	3	708	3	848	3	1010	3	1204	3	1444
4	587	4	711	4	851	4	1014	4	1208	4	1449
5	589	5	713	5	854	5	1018	5	1213	5	1455
6	591	6	716	6	857	6	1021	6	1217	6	1460
7	594	7	718	7	860	7	1025	7	1222	7	1466
8	596	8	721	8	863	8	1028	8	1226	8	1471
9	598	9	724	9	866	9	1032	9	1230	9	1477
31:0	601	36:0	726	41:0	869	46:0	1035	51:0	1235	56:0	1482
1	603	1	729	1	872	1	1039	1	1239	1	1488
2	606	2	732	2	875	2	1043	2	1244	2	1494
3	608	3	734	3	878	3	1046	3	1248	3	1499
4	610	4	737	4	882	4	1050	4	1253	4	1505
5	613	5	740	5	885	5	1054	5	1257	5	1511
6	615	6	743	6	888	6	1057	6	1262	6	1516
7	618	7	745	7	891	7	1061	7	1266	7	1522
8	620	8	748	8	894	8	1065	8	1271	8	1528
9	622	9	751	9	897	9	1069	9	1275	9	1534
32:0	625	37:0	753	42:0	900	47:0	1072	52:0	1280	57:0	1540
1	627	1	756	1	903	1	1076	1	1284	1	1546
2	630	2	759	2	907	2	1080	2	1289	2	1552
3	632	3	762	3	910	3	1084	3	1294	3	1558
4	635	4	764	4	913	4	1087	4	1298	4	1564
5	637	5	767	5	916	5	1091	5	1303	5	1570
6	639	6	770	6	919	6	1095	6	1308	6	1576
7	642	7	773	7	923	7	1099	7	1313	7	1582
8	644	8	776	8	926	8	1103	8	1317	8	1588
9	647	9	778	9	929	9	1107	9	1322	9	1594
33:0	649	38:0	781	43:0	932	48:0	1111	53:0	1327	58:0	1600
1	652	1	784	1	936	1	1114	1	1332	1	1606
2	654	2	787	2	939	2	1118	2	1337	2	1613
3	657	3	790	3	942	3	1122	3	1342	3	1619
4	659	4	792	4	946	4	1126	4	1346	4	1625
5	662	5	795	5	949	5	1130	5	1351	5	1632
6	664	6	798	6	952	6	1134	6	1356	6	1638
7	667	7	801	7	956	7	1138	7	1361	7	1645
8	669	8	804	8	959	8	1142	8	1366	8	1651
9	672	9	807	9	962	9	1146	9	1371	9	1658
34:0	674	39:0	810	44:0	966	49:0	1150	54:0	1376	59:0	1664
1	677	1	813	1	969	1	1154	1	1381	1	1671
2	679	2	815	2	972	2	1158	2	1386	2	1677
3	682	3	818	3	976	3	1163	3	1392	3	1684
4	685	4	821	4	979	4	1167	4	1397	4	1691
5	687	5	824	5	983	5	1171	5	1402	5	1698
6	690	6	827	6	986	6	1175	6	1407	6	1704
7	692	7	830	7	989	7	1179	7	1412	7	1711
8	695	8	833	8	993	8	1183	8	1417	8	1718
9	698	9	836	9	996	9	1187	9	1423	9	1725
35:0	700	40:0	839	45:0	1000	50:0	1192	55:0	1428	60:0	1732

D. T.	¼ part of the Girt.
0:0	I. T.
1	0:1
2	0:3
3	0:5
4	0:7
5	0:9
6	1:0
7	1:1
8	1:3
9	1:4
1:0	1:5
1	1:7
2	1:9
3	2:1
4	2:3
5	2:4
6	2:6
7	2:7
8	2:9
9	3:1
2:0	3:3
1	3:5
2	3:6
3	3:8
4	4:0
5	4:1
6	4:2
7	4:4
8	4:6
9	4:7
3:0	4:8
1	5:0
2	5:2
3	5:4
4	5:5
5	5:7
6	5:9
7	6:1
8	6:2
9	6:4
4:0	6:6
1	6:8
2	6:9
3	7:1
4	7:3
5	7:5
6	7:6
7	7:7
8	7:9
9	8:0
5:0	8:2

A TABLE

Shewing the Fourth Part of the Girt in Inches and Tenths, anſwering to every Tenth Part of a Degree, ſo far as ten Degrees.

The USE of theſe TABLES will appear very plain from one Example.

Meaſure from the Tree, ten, twenty, thirty, &c. Feet, and there plant the Theodolite level.

Direct the Teleſcope to the Bottom of the Tree, and obſerve the Degree and Tenth of Depreſſion, and to the Top of the Tree, the Degree and Tenth of Elevation.

Then find in the Table, the Parts anſwering to each, which being added together, make the exact Height of the Tree.

But, becauſe the Fourth Part of the Girt in Inches muſt be taken in the Middle of the Tree; ſubtract the Parts anſwering to the Depreſſion, from half the Height of the Tree: The Remainder ſeek for in the Table, under Parts, againſt which, in the left hand Column, is the Degree and Tenth of Elevation; to which, if you ſet the Teleſcope, and obſerve the Diameter of the Tree, i. e. the Degree and Tenth on the Limb, anſwering to the Diameter, and bring them to this Table, you will have the Fourth Part of the Girt in Inches, and Tenths anſwering thereto.

	D. T.	Parts.
Depreſſion.	24:3	451
Elevation.	58:8	1651
Height of the Tree.	. . . .	21.02
Half the Height. . . .	. . . .	10.51
		451
Set the Teleſcope. .	31:0	600

		¼ part of the Girt.
	D. T.	I. T.
Diameter on the Limb.	9:6	16:0

N. B. Theſe Tables are calculated for the Diſtance of ten Feet; yet are they as perfect for twenty, thirty, &c. Feet, only by doubling, trebling, &c. the Length, Diameter, and Fourth Part of the Girt found thereby.

D. T.	¼ part of the Girt.
5:0	I. T. 8:2
1	8:4
2	8:6
3	8:8
4	9:0
5	9:1
6	9:3
7	9:4
8	9:6
9	9:8
6:0	10:0
1	10:2
2	10:4
3	10:5
4	10:7
5	10:8
6	10:9
7	11:0
8	11:2
9	11:4
7:0	11:6
1	11:8
2	12:0
3	12:1
4	12:3
5	12:5
6	12:6
7	12:8
8	13:0
9	13:1
8:0	13:2
1	13:4
2	13:6
3	13:8
4	13:9
5	14:0
6	14:2
7	14:4
8	14:6
9	14:8
9:0	14:9
1	15:0
2	15:2
3	15:4
4	15:5
5	15:7
6	16:0
7	16:1
8	16:2
9	16:3
10:0	16:6

NOTES
ON THE
Practical Surveyor

PREFACE

SOme may queſtion the Utility of reprinting a ſurveying Book of this Age, but as Mr. Wyld argues in his Preface, the Arts Mathematical can never be fully learned. Much Time has paſſed, and many of theſe Methods and Inſtruments are on the verge of being forgotten, yet they ſtill are convenient and of intereſt to the Practiſer of the Art.

The Editor eſpecially thanks R.R.M. for her comments and enthuſiaſm.

David Manthey

SECT. I.

Deſcribing the Printing Hiſtory, Differences in the Firſt Edition, the current Typeſetting, &c.

SAMuel Wyld's Book has at leaſt the following printing Hiſtory:

[1] Firſt Edition, 1725. pp. xv. 182. 1 front Plate, 5 end Plates.
[2] Second Edition, 1730?. pp. viii. 188. 6 or 7 Plates.

Plates.

[4] Fourth Edition, 1760. pp. viii. 191. 1 front
     Plate, 5 Plates placed between Chapters.

[5] Fifth Edition, 1764. pp. viii. 191. 6 Plates.

[6] Sixth Edition, 1769 or 1770.

[7] Seventh Edition, 1780. pp. viii. 191.

The text of this First Edition has been taken
from a Microfilm produced by Research Publica-
tions, being a Print of a Volume found in the
British Museum, and with the Microfilm located
at the New York State Library. An original Volume of the First Edition was also examined in the
Boston Public Library. There are differences
between the title Pages of the British Museum
and the Boston Public Library copies. That presented as the title Page is a faithful Copy of the
Volume from the British Museum.

Based on an Examination of different Editions
of the Book, it is judged that the Volume from
the Boston Public Library is from an earlier
Printing than that from the British Museum.

On the following Page is the title Page as
found in the Volume from the Boston Public
Library:

THE

THE
Practical Surveyor,

OR, THE

Art of *Land-Measuring*,

Made E A S Y.

Shewing by plain and practical Rules, how to Survey any Piece of Land whatfoever, by the Plain-Table, Theodolite, or Circumferentor: Or, by the Chain only. And how to Protract, Caft up, Reduce, and Divide the fame.

L I K E W I S E,

How to Protract Obfervations made with the Needle: and how to Caft up the Content of any Plott of Land: By Methods more Exact and Expeditious than heretofore ufed.

To which is added,

An A P P E N D I X,

Shewing how to Draw the Plan of Buildings, &c. in Perfpective, from Obfervations made by the Theodolite. As alfo the Ufe of a new-invented *Spirit-Level*. With feveral other Things never before made Publick.

Socrates, *hunc finem Geometrie Principalem effe ftatuebat; Ut agrum planum metire dividereq; poffit.*

Pitifcus Geod.

L O N D O N:

Printed for J. H O O K E, at the *Flower-de-luce* againft St. *Dunftan's* Church in *Fleet-ftreet*: And J. S I S S O N, Mathematical Inftrument-maker, the Corner of *Beaufort-Buildings* in the *Strand*. M,DCC,XXV.

Although the Text of this Volume is meant to be an exact and perfect Replica, there are fome differences in Formatting and layout of the printed Page. Specifically, fome of the word Spacing is different, and alfo fome of the Pagination. Additionally, many of the Tables have been fquared and aligned, as the original Printing was fomewhat offset in Parts. In all Cafes, the Spelling is that of the Original, be it ever fo Varied, excepting where the Typefetter obvioufly made a Miftake, fuch as inverting the Letter *u*, or placing a double Comma.

The Font ufed in this Copy was developed for this Printing. It is bafed on the Caflon Font, with many Letters mildly or fubftantially altered and Spacing much improv'd.

SECT. II.

Commentary on the Book

THE title Page prefented in the Notes contains the Latin Phrafe: Socrates, *hunc finem Geometrie Principalem effe ftatuebat; Ut agrum planum metire dividereq; poffit*; Which tranflates to Englifh: *Socrates confiders this to be the firft Purpofe of Geometry: to be able to Meafure and Divide a flat Field.* This Phrafe is credited to Bartholomeo Pitifcus, being the Author of mathematical Works on Geodefy and Trigonometry; Pitifcus was the firft Perfon to ufe the term Trigonometry.

In the Preface, there is the Latin Phrafe: *Ornari res ipfa negat contenta doceri*; Which tranflates to Englifh: *The Thing itfelf refufes to be Commended; it is fufficient that it is Taught.*

This is followed later by the Phrafe: *Rumpatur quifquis rumpitur invidiâ*; Which tranflates

flates to Englifh: *Whoever is broken by envy, let him be broken.*

At the End of the Preface is the French Phrafe: *Va, mon Enfant, prend ta Fortune*; Which tranflates to Englifh: *Go, my Child, feek your Fortune.*

In the Table of Contents, Chap. II., Sect. 6. is not lifted. This matches the firft Printing.

In Chap. I, Sect. 1, prior to Edward I of England, an Acre was the amount of Land a Yoke of Oxen could plow in a Day. Edward I decreed that henceforth an Acre fhould always contain 160 fquare Perches.

In the *Table of Square Meafure*, the original Printing incorrectly afferts that 1 fquare Yard contains 20.755 fquare Links; in this Printing, it has been corrected; 1 fquare Yard contains 20.661 fquare Links. Interestingly all of the Tables in Sect. 1, are identical to Tables found in the Eighth Edition of *Geodæfia*, a Book publifhed in 1768, and originally written by John Love in 1688; *Geodæfia* alfo contains the incorrect Number of fquare Links in a fquare Yard.

In Chap. IV, in the fourth Page of Sect. 2, the original Printing reads: *both denoting the North-weft Corner of the* Stockin; however, bafed on the Plott in *Fig.* 22., this fhould indicate the *South-weft Corner*; as a confequence, the Text has been changed to read South-weft.

In Chap. V, the method of meafuring Angles ufing a Chain is quite exact. If one meafures the Diftance a, b, to within $\frac{1}{10}$ of a Link of its true Length, this yields an Angle that is accurate to flightly better than ± 2 Minutes; Additionally, the Meafurement is flightly more exact for fmall Angles than for large Angles.

Alfo in the 5th Chapter, in the 3d Section there

there is an Equivalence given between an Angle
whofe Quantity is expreffed in Degrees and
Minutes and that fame Angle with the Quantity
expreffed in Sextants and Parts. In the original
Text, the Angle 102°, 20', is noted as 1 Sextant
734 Parts, and alfo the Quantity 230°, 50' is given
as 3^S, 886^P; yet ufing an accurate Table of Sines,
the Number of Parts is different.

To accurately calculate the number of
Sextants and Parts from an Angle, firft divide the
Quantity of the Angle in Degrees by 60, the
Number of Degrees in a Sextant; this will give
the Number of Sextants. Next, fubtract from
the Quantity of Degrees, the Number of Sex-
tants multiplied by 60; taking the Remainder,
along with the Minutes, divide it in half, remem-
bering that 1 Degree is equal to 60 Minutes.
Now, look in a Table of Sines for the Refult of
this Divifion, and taking the Number from the
Table, double it. Laftly, multiply the doubled
Number by 1000, difcard any Remainder; this
Figure is the Number of Parts.

As an Example, to convert the Angle with a
Quantity of 102 Degrees, 20 Minutes to a Quantity of Sextants and Parts, firft divide the Degrees
by 60, giving 1 Sextant and fome inconfequential
Remainder. From 102 Degrees, fubtract the
Number of Sextants 1, multiplied by 60, or in
this Cafe the number 60, leaving 42 Degrees, 20
Minutes; and then take the half of this Value, 21
Degrees, 10 Minutes. From a Table of Sines, it
is found that 21 Degrees, 10 Minutes correfponds
to a Sine of $\frac{36108}{100000}$, which is firft doubled to $\frac{72216}{100000}$,
then multiplied by 1000, giving 722 and $\frac{16}{100}$;
whereupon the Fraction is difcarded, leaving 722
Parts. Therefore, the Quantity of 102 Degrees,
20 Minutes is equivalent to a Quantity of 1
                                                    Sextant

Sextant and 722 Parts.

In like Manner, the Quantity of Degrees and Minutes can be found given the Quantity of Sextants and Parts; Take the number of Parts and divide it by 1000, then take the half of this; find the clofeft match in the Table of Sines and read the correfponding Number of Degrees and Minutes, and double it, again recollecting that there are 60 Minutes in each Degree. To the Quantity of Degrees, add the Number of Sextants multiplied by 60. For example, if there are 3 Sextants, 858 Parts, take the Number of Parts 858, divide it by 1000 to get $\frac{858}{1000}$, then take the Half of this, which is $\frac{429}{1000}$. Confulting the Sine Table, the clofeft Value found is $\frac{42920}{100000}$ immediately next to the Quantity 25 Degrees, 25 Minutes; doubling this Quantity gives 50 Degrees, 50 Minutes. Laftly, taking the Number of Sextants 3, and multiplying by 60, gives 180; this is added to the Degrees and Minutes, giving 230 Degrees, 50 Minutes.

In Chap. VI, Sect. 2, there is the Phrafe *mutatis mutandis*, which tranflates to *with fuitable or neceffary Alterations*.

In Chap. VII, Sect. 3, moft of the Text is original, but a few Sentences are copied verbatim from *Geodæfia* by John Love; or, perhaps, *vice verfa*.

In the fecond Section of the firft Chapter in the Appendix, there is a Table of the Earth's Curvature: Bafed on this Table, the Radius of the Earth is between 3973 Miles and 4005 Miles; The currently accepted Equatorial Radius is 3963 Miles. It is likely that the Table was calculated bafed on the Rule given in the fourth Section.

In the fourth Section, a Formula is given for calculating the Effect of Curvature; this Formula

is

is the firſt order Taylor-ſeries Expanſion of the Radius of Curvature Equation, aſſuming that the Radius of the Earth is 3992 Miles; Although the Rule is not exact, it is within a Thouſandth of a Foot for more than 1000 Chains. Uſing a Radius of 3966 (which is much cloſer to the current accepted Value), the Formula is changed to this Rule: Multiply the Square of the Diſtance in Chains by 104, and divide the Product by 1000000.

In Chap. 3, Prob. 1. of the Appendix, it ſtates that to find the greateſt Angle on the Weſt of the Meridian, add 11 h. 84 m. to the Time found in the Table. This is correct; it would more commonly be ſaid to add 12 h. 24 m. to the Time.

The Two Tables for uſe with the *Pole Star*, are only valid for a limited Number of Years, due to the movement of the Earth and the reſpective Motion of the Stars. Baſed on the *Aſtronomical Ephemeris Calculator* made by *S. L. Moſhier*, theſe Tables appear to be formulated for the Year 1656 (give or take ſome few Years); While only minor Changes occur each Year, after a ſignificant Number of Decades, the Tables are no longer appropriate. A new ſet of Tables is preſented in Sect. 3. of theſe *Notes*.

Mr. *Moſhier*'s Aſtronomical Ephemeris Calculator is available at many Locations, along with compleat Details on how his Computations are performed; a ſearch of a global Index ſhould ſhew where to obtain the Calculator.

S E C T.

SECT. III.

Tables for finding the Meridian using the Pole Star for the Year MMI.

HERein are Two Tables, as presented in the First Problem of the Third Chapter of the Appendix; they have been adjusted for a new Year. The Tables are used in the same Manner as previously directed.

The following Table shews the Time from Noon, when the Pole Star makes the greatest Angle from the Meridian-Line towards the East, for any Time in the Year.

January.			February.			March.			April.		
D.	H.	M.	D.	H.	M.	D.	H.	M.	D.	H.	M.
1	1	51	5	23	33	5	21	42	2	19	52
8	1	24	12	23	5	12	21	14	9	19	24
15	0	56	19	22	37	19	20	47	16	18	56
22	0	28	26	22	10	26	20	19	23	18	29
29	0	0	—			—			30	18	1

May.			June.			July.			August.		
7	17	34	4	15	44	2	13	27	6	11	34
14	17	7	11	15	17	9	13	0	13	11	7
21	16	39	18	14	49	16	12	33	20	10	39
28	16	11	25	14	22	23	12	4	27	10	12
—			—			30	11	49	—		

September.			October.			November.			December.		
3	9	45	1	7	55	5	5	38	3	3	48
10	9	17	8	7	28	12	5	11	10	3	20
17	8	50	15	7	1	19	4	43	17	2	53
24	8	23	22	6	33	26	4	15	24	2	25
—			29	6	6	—			31	1	57

The

The preceding Table fhews the Time from Noon, where Noon is the mean Solar Time; Suppofe the Meridian is to be found on the firft of October, then the Pole Star will be at its greateft Angle to the Eaft of the Meridian 8 Hours and 5 Minutes after Noon.

The following Table fhews the greateft Angle which the Pole Star makes with the Meridian in any of thefe Latitudes, (viz.)

Latitude.		Angles.		Latitude.		Angles.	
Deg.	Min.	Deg.	Min.	Deg.	Min.	Deg.	Min.
30	00	0	51	45	00	1	02
31	00	0	51	46	00	1	04
32	00	0	52	47	00	1	05
33	00	0	53	48	00	1	06
34	00	0	53	49	00	1	07
35	00	0	54	50	00	1	09
36	00	0	55	51	00	1	10
37	00	0	55	52	00	1	12
38	00	0	56	53	00	1	13
39	00	0	57	54	00	1	15
40	00	0	58	55	00	1	17
41	00	0	59	56	00	1	19
42	00	0	59	57	00	1	21
43	00	1	00	58	00	1	23
44	00	1	01	59	00	1	26

The fame Method of obferving the Sun before and after Noon can alfo be ufed on any cæleftial Object before and after it croffes the Meridian; The Pole Star croffes the Meridian twice each Day, approximately 6 Hours before and 6 Hours after the Time lifted in the preceding Table.

S E C T.

SECT. IV.

Being a Lift of all Inftruments and Tools mention'd in the Book.

ARRows – Sticks about a Foot long with a red Cloath at the top and a Iron Ferrill at the bottom; Nine are us'd together with the Chain.

Beam-compaffes – Two Pins or drawing Points are mounted on a brafs Bar; one of thefe Pins can be mov'd forward and back fo as to be at any Diftance apart from the other Pin. The Pins are ufed in the fame manner as the Points of a regular Pair of Compaffes. Often the Beam is fcribed with a Scale.

Bevel – An Inftrument with two Rulers, jointed together at one End, and opening to any Angle.

Brafs Arch – A fmall Arch of a Circle, with a Hair ftrung between fmall Holes drill'd near the ends of the Arch. It is ufed when cafting up a Plott to eftimate the average bounder Line along Hedges or other fmall Features.

Chain – A *Gunter's* Chain is 4 Poles, where each Pole is exactly 16 $\frac{1}{2}$ Feet in Length, as *per* Statute. The Chain is divided into 100 equal Links, each Link being 7 $\frac{92}{100}$ Inches, with a brafs Marker at every tenth Link. A red Cloth is tied at 50 Links, and others of a lighter Colour at 25 from each end of the Chain.

A Chain which is 2 Poles in Length is more convenient for meafuring ploughed Lands in Common-Fields; this is identical to the *Gunter's* Chain, excepting that there are only 50 Links. When meafuring Cities or Towns, a Chain confifting of Links that are of a Foot

long

long is more practical, as Towns are generally laid out by Foot Measure; 'tis useful for such a Chain to consist of Fifty Links, being 50 Feet in Length.

Circumferentor – Otherwise known as a Box-Compass, this is a Box with a magnetick Needle which can be mounted on a Staff or Tripod by use of a Ball and Socket; a Limb gives the Angle that the Needle forms with the Sights. The Sights on the Box can either be a plain Index or a Tellescope.

Compasses – A Pair of Arms, each with a Pin or black-lead Tip used for drawing Arcs; the Arms are hinged firmly at the Top and can be open'd any amount.

Cross – This can be made in several divers Forms, all of which serve to the same Purpose. The Cross is used to find a Point along a Right Line such that the Right Line from the Point to a distant Mark stands at Right Angles to the base Line. It can either be made from a Cross with four plain Sights, from a Board with four Pins, or from a brass Cylinder with four pierc'd Slots; in all Cases, one pair of Sights must be perpendicular to the Other, and the Cross is designed to be placed on a Staff.

Drawing Board – A Table that is ruled with a Scale on each Edge. This is usually us'd with a Tee to facilitate drawing Lines.

Drawing Pen – A Pen used to draw an *Indian*-Ink Line; the Line should be very narrow.

Field-Book – A note Book wherein all Measurements and Observations of a Survey are written so that later the Work may be protracted and truly cast up.

Index – A strait Ruler with Sights on either end, used with the Plain-Table to draw Right Lines

in

in the Direction of a Station or other Feature; the Ruler has a fiducial Edge that is optionally marked with a Scale, and the Sights have a vertical Hair or are made as very narrow Slits.

Needle – A magnetick Needle mounted in a Box for ufe with the Plain-Table. This alfo refers to the Needle that is Part of the Theodolite. The Box has a Card mark'd with Degrees and a Flower-de-luce. The Needle refts on a Pin, and is often fecur'd with a Cap.

Offset-Staff – A Staff whofe Length is equal to $\frac{1}{10}$ part of the Chain, and is mark'd every Link, with one Link being divided into Tenths.

When furveying Cities and Towns, it is ufeful to have a Staff that is Five Feet in Length, and is mark'd every Foot, with one Foot fubdivided; alfo an additional Piece of about 3 Feet in Length that can be joined to make a Square at one end of the Staff is of utility in laying off Offsets.

Paddle – A fmall Spade or other Tool ufed for cutting the Ground or Turf to mark where a Station was located.

Parallelogram – An Inftrument ufed to either reduce or enlarge a Plott. See the front Plate for a Drawing of a Parallelogram.

Parallel Ruler – A Pair of ftrait Pieces of Brafs or other Material, join'd fo that they can be opened to any Diftance, and fo that the Edges of the Pieces always remain parallel.

Pins – Ufed in plotting and drawing on the Plain-Table, thefe can be ufed to lightly fcribe the Paper or to mark a Point.

Plain-Table – A Table mounted on a Swivel that can be placed on top of a Ball and Socket on a Tripod. The Table has a Frame to hold a piece of Paper or Parchment fecurely, and this

Frame

Frame is generally mark'd with a Scale. The Plain Table is ufed with an Index to make a Plott directly in the Field.

Plummet – A Plumb-bob on a String which can be hung from a Tripod to place an Inftrument directly over a Hole or Mark on the Ground.

Protractor – Although a Protractor that is a Semicircle can be ufed, a whole Circle is better. In general the Protractor is an Inftrument for meafuring Degrees and Minutes. An Inftrument for meafuring Sextants and Parts, or Sextants and Links, can alfo be made.

Both Mr. Ward's Protractor and Mr. Siffon's Protractor have an Index that can be moved. The Index of Mr. Ward's Protractor has a ftreight Edge with irregularly fpaced Marks indicating the Minutes of the Arch. The Index of the Protractor is fhewn in the *Figure* below.

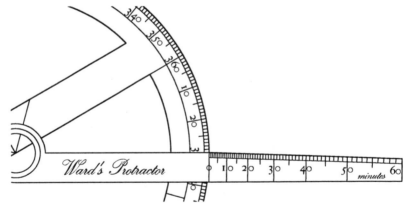

Ward's Protractor

The Index of Mr. Siffon's Protractor has a curved Edge with uniformly fpaced Marks giving the Minutes of the Arch. The curved Edge of the Index has a Radius fuch that a Chord of 2 Degrees is the fame Length as the mark'd part of the Index, and a Continuation of this Arch will pafs through the exact Center of of the Protractor. The Index of the Protractor is ſhewn in the following *Figure*.

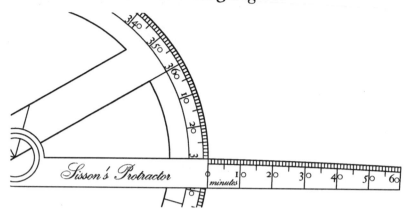

The ſame type of Protractor can be made without a movable Index, ſtill allowing the Inſtrument to meaſure Minutes. In this Caſe, there is a Limb cut into the Croſs-Bar of the Protractor; the Limb is an Arch ſimilar to the Arch on the Index in the preceding Example. Since the Minute Limb is much leſs in Extent on the Protractor without an Index, it is not as accurate; however there are no moving Parts, ſo 'tis impoſſible for the Protractor to ſhake or ſlip. A partial Drawing of ſuch a Protractor is ſhewn. See the following *Figure*.

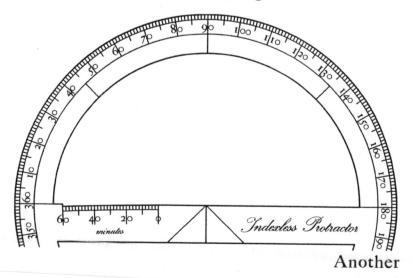

Another

Another Inſtrument for protracting the Degrees and Minutes is the Vernier Protractor. The Index of this Protractor is a Right Line, often mark'd with a Scale, and with a ſmall Limb with Vernier Markings. Unleſs the Limb of the Index be very large, the Marks are uſually every 2 Minutes. The Quantity of Minutes in an Angle are found on the Vernier Scale on the Index; the Mark that is beſt align'd with a Degree Mark on the Limb of the Protractor ſhews the Minutes of the Angle. The Index of a Vernier Protractor is ſhewn in the following *Figure.*

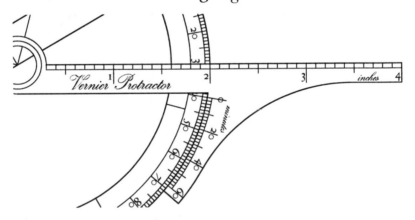

When protracting Meaſurements made with a Circumferentor, it may be uſeful to mark the Location on the Protractor where 360 Degrees is mark'd on the Limb with a Flower de luce or with the Letter N or North, and the Location where 180 Degrees is mark'd with the Letter S or South; this will aſſiſt in keeping the Protractor in its true Poſition.

When protracting Angles that have been meaſured uſing the Chain, the Angles can be drawn with a Scale and Compaſſes. It may be more convenient to have a Protractor that is mark'd in Sextants and Links. Such a Protractor

tractor is fhown in the *Figure* below.

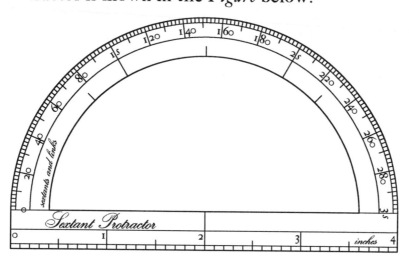

Quadrant – A fmall Tool ufed when meafuring
with the Chain. This is ufed to meafure the
Quantity of the Angle of Altitude on an Hill;
It is fimilar to a Gunner's Quadrant, being a
Sight with a graduated Limb below it, and a
Plummet and String attached fo that the
String cuts the Limb to indicate the vertical
Angle.

Scale – Ufed in plotting, the Scale is decimally
divided clofe to the Edge; diagonal Lines can
be made acrofs the Scale allowing a Meafurement to be made to within $\frac{1}{100}$ of an Inch. A
plotting Scale has a champhered Edge with
the Markings running to the Edge. See the
Figure below.

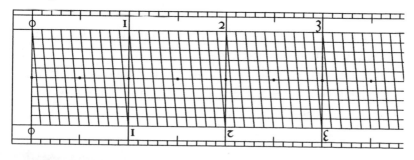

To meaſure a Diſtance on the Scale, ſuppoſe 2 $\frac{37}{100}$ inches, find the 3d diagonal Line to the right of the 2 Inch Mark, then find the 7th horizontal Diviſion below the Top of the Scale. Place one End of the Compaſſes where the diagonal Line croſſes the horizontal Line, and place the other Foot of the Compaſſes at the 7th horizontal Diviſion along the Line mark'd o (zero) at the left Edge of the Scale.

Sea-Compaſs – Alſo called a Mariner's Compaſs, this is a Box with a magnetick Needle mounted therein, the entire Apparatus fixed to a Gimbal ſo as to always remain upright no matter the Motion of the Seas; Rather than having the Limb divided in 360 Degrees, the Sea-Compaſs is divided into 32 Points.

Sector – Two Rulers connected at one End by a Joint, each of which is marked with ſeveral Scales, with a matching Scale of each kind on each Arm. The Scales begin at the Center of the Joint.

Spirit-Level – A Spirit-Tube with a Bubble that is centered when the Tube is exactly level, fitted to a Telleſcope and mounted on a three-legged Staff. See the Drawing on the front Plate.

Square – Uſed for drawing Perpendiculars. This is an optional Tool, as the ſame Task can be accompliſhed uſing only Compaſſes and a Ruler.

Station-Staff – A Staff or Rod placed at a Station where the Theodolite was erected.

When uſed for Levelling, the Station-Staves ſhould be 10 Foot long and divided into 1000 equal Parts; each Staff is fitted with a Vane which can be adjuſted to ſhow a Diviſion on the Staff with the utmoſt exactneſs. It is

ſometimes

ſometimes neceſſary to add to the Staff an additional Piece 5 Foot in Length and divided into 500 equal Parts.

Staves – Uſed for marking Bounders of Fields and for meaſuring with the Chain; each Staff ſhould be ſtrait and about 5 Feet long.

Steel-Bow – An Inſtrument uſed for drawing curved Lines, fitted with Screws to adjuſt the Shape of the Curve.

Tee – A T-Square for uſe with a Drawing-board.

Theodolite – An Inſtrument that can be placed level and can meaſure both horizontal and vertical Angles.   See the firſt Section of the ſecond Chapter.

Wheel – Alſo called a Way-wiſer. This is a round Wheel that is run along the Road; each time the Wheel revolves, it moves a Hand on a Limb which points to the Number of Poles and Furlongs that have been travelled.

SECT. V.

Being a Gloſſary of Words that the Editor found either Hard or of intereſt to the Publick.

ALLom – Alum.   Principally Aluminum Sulfate or one of its Salts. Potaſſium Alum is produced by treating Bauxite with ſulfuric Acid and then with Potaſſium Sulfate.

Azimuth – The Horizontal Angle from the true Meridian.

Azure – Pigment made from Lapis Lazuli.

Black-lead – Graphite.

Caſt-up – To calculate the Area and other ſalient Characteriſtics.

Co-latitude – The Latitude ſubtracted from 90°.

Cypher – The Numeral Zero. When a Cypher is annexed to a Number, it is the ſame as
multiplying

multiplying that Number by Ten.

Demefnes – Legal parcels of Land.

Ephemeris - A Table lifting the Places of the cæleftial Bodies for each Day of the Year.

Epitome – A compact or condenfed Reprefentation of Something.

Flower-de-luce – The *fleur-de-lis*, typically ufed as an Ornament for an Arrow pointing North. Named after a group of Flowers having fimilar Leaves.

Gum-Water – Gum of the Acacia diffolved in Water.

Indian-Ink – Lamp-black or Ivory-black, and animal Glue. Alternately, a Pigment made from the fecretion of the Cuttlefifh and treated with cauftic Potafh; this is alfo known as Sepia.

Indico – A Color extracted and proceffed from the various Species of the *Indigofera* plant, made by fteeping the Leaves and Stalks, then agitating the Tincture 'til the Dye begins to form Granules; thefe Granules are dried and ufed as a Pigment.

Meridian-Line – A line of Longitude; that is, a line running true North to true South.

Nonus's Invention – A method for meafuring Angles to an Accuracy greater than a Degree. It was invented by the Portuguefe aftronomer Pedro Nuñes in 1542. His Defcription follows (as taken from a tranflated Source).

> *drawing on the face of a Quadrant for meafuring Angles 45 concentric Arcs; the outfide Arc is divided into 90 equal Parts or Degrees, and the remainder into 89, 88, 87, 86, &c. fucceffively, the Laft being divided into 46 equal Parts. When the Index does not exactly cut one of the Divifions of the Arc of Degrees, it paffes*
> *through*

through or near a Division on an other Arc; and by noting the Place of that Division the Parts of a Degree are calculated.

It is more likely that Vernier's Scale is used, it being simpler and more accurate; see Sect. 4 for a Drawing of a Vernier.

Obscure – Indistinct or hidden. An obscure Line should be very faint, as it is used for the construction of a Plott, and is not part of the Bounders or other Point of Interest.

Occult – Hidden. An occult Line should be remov'd from a Plott after the Bounders have been drawn.

Oker – Ocher. A ferruginous Clay used as a pigment; usually this refers to a Limonite Clay.

Okers brown of *Spain* – Probably Hematite-based Ocher.

Pales – A picket Fence or a Rail supported by Pickets.

Panegyrical – Laudatory.

Pink – Chrome Aluminum Stannate.

Pounce – A fine Powder, typically from sandarac or cuttlefish Bone, rubbed on the Paper to prevent color Bleed. The Act of rubbing such a Powder on Paper.

Protract – To draw to scale or to plot.

Pumiced – To clean or smooth with Pumice.

Quickset Hedge – A Hedge grown from Cuttings that have been planted, usually of Hawthorne.

Rosin – Pine Resin.

Rood – An area equal to $\frac{1}{4}$ of an Acre.

Sadder – Dark or drab.

Searced – Sifted or bolted.

Smalts – A deep Blue made from pulverizing blue Glass made from Silica, Potash, and
Cobalt

Cobalt Oxide.

Sough – A Drain or Adit.

Stanifh-Grain – Poffibly a tin-bafed Powder.

Superficies – A Surface; the exterior Part or Area.

Terrier – A Lift in which Lands are defcribed by their Site, Bounders, Quantity of Acres, and the like.

Trencher – A fquare Plate. Alternately, a mortar-board Hat.

Umber – Brown or reddifh Pigment made from Clay containing Iron and manganefe Oxide.

Verdigreafe – The greenifh copper Oxide formed on Copper; fpecifically, Copper Carbonate.

White – Probably White-lead.

White-lead – Lead Carbonate.

FIGURES,

TO ACCOMPANY THE TEXT.

The original Copy of *The Practical Surveyor* had Six foldout Plates, One of them fronting the Title Page, and the other Five Plates at the Back of the Volume; each of thefe was printed from Copper-Plates and meafured $11\frac{1}{4}$ by 9 Inches in Size. Thefe Plates are reproduced herein, tho' they are each fplit between Two Pages.

Fig: 1.

Fig: 2.

Fig: 3.

Fig: 4.

Fig: 5.

Fig: 6.

Fig: 7.

Fig: 8.

Fig: 9.

Fig: 10.

Fig: 11.

Fig: 12.

Fig: 13.

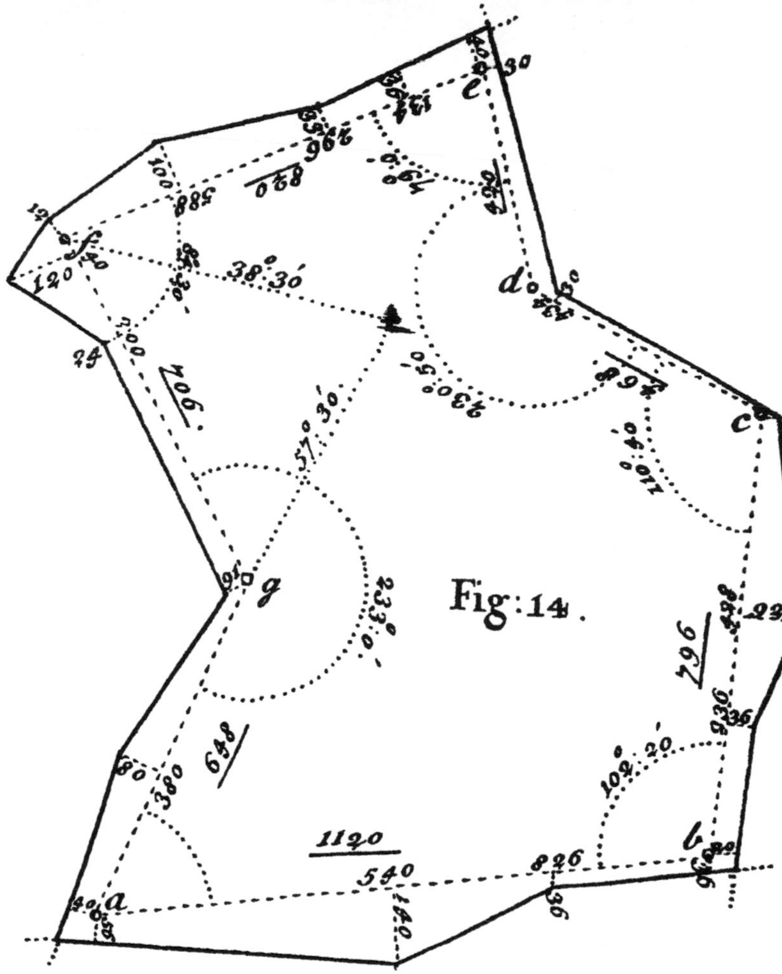

Fig: 14.

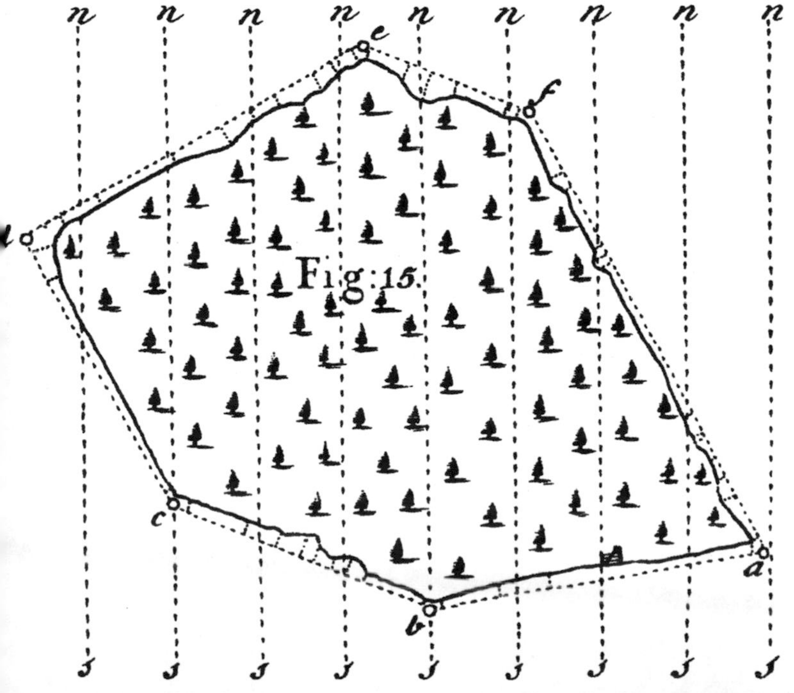

Fig: 15.

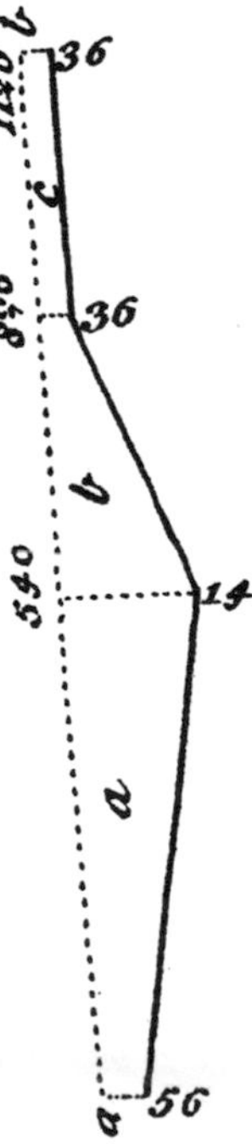

Fig: 16.

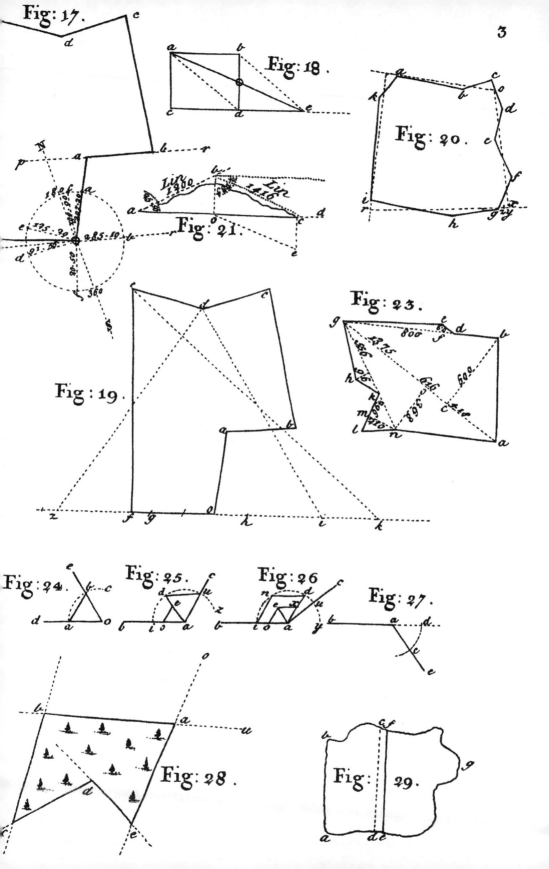

Fig: 17.

Fig: 18.

Fig: 20.

Fig: 21.

Fig: 19.

Fig: 23.

Fig: 24.

Fig: 25.

Fig: 26.

Fig: 27.

Fig: 28.

Fig: 29.

Fig: 22.

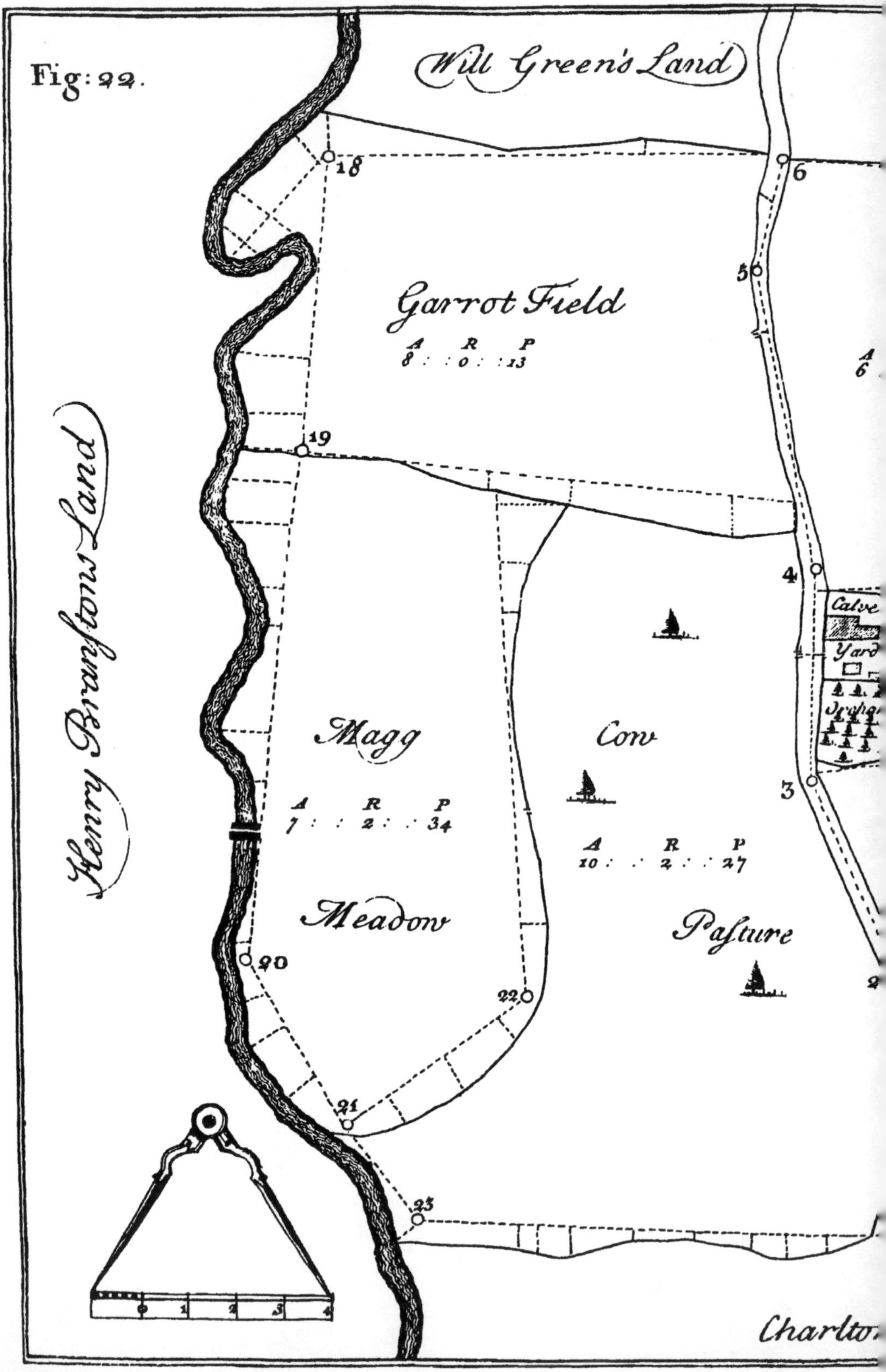

Will Green's Land

Henry Branftons Land

18

6

5

Garrot Field
A R P
8 : 0 : 13

19

4

Calve
Yard

Magg
A R P
7 : 2 : 34

Cow
A R P
10 : 2 : 27

3

Meadow

20

22

Pasture

2

21

23

Charlto.

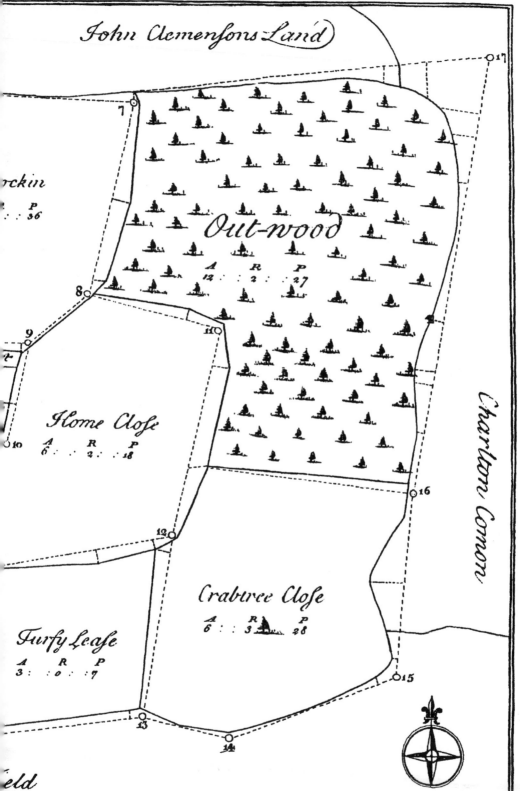

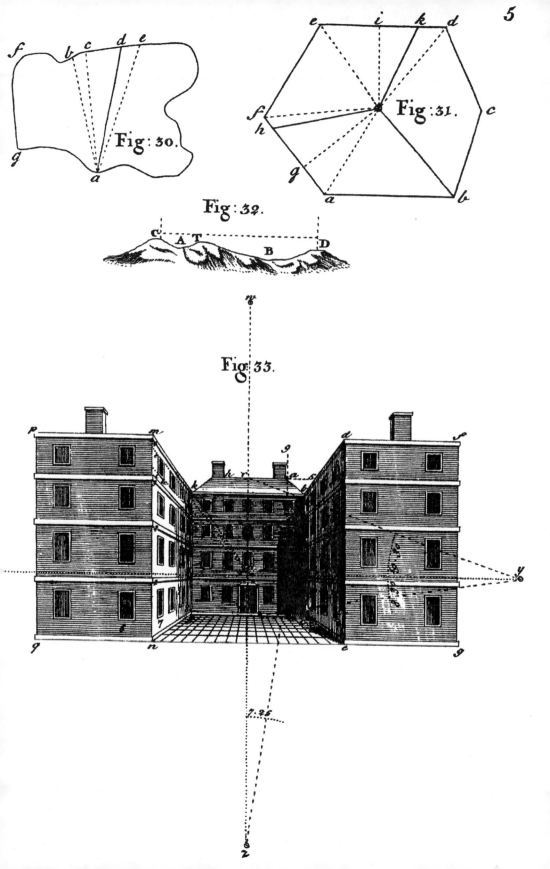

Fig: 30.

Fig: 31.

Fig: 32.

Fig: 33.

5

Printed in the United States
38347LVS00005B/64